The Greenhouse Effect, Climate Change, and U.S. Forests

The Conservation Foundation is a nonprofit research and communications organization dedicated to encouraging human conduct to sustain and enrich life on earth. Since its founding in 1948, it has attempted to provide intellectual leadership in the cause of wise management of the earth's resources. The Conservation Foundation is affiliated with World Wildlife Fund.

The Greenhouse Effect, Climate Change, and U.S. Forests

Edited by
William E. Shands and
John S. Hoffman

The Conservation Foundation
1250 Twenty-fourth Street, N.W.
Washington, D.C. 20037

The Greenhouse Effect, Climate Change, and U.S. Forests

Cover design by Patrick Murphy
Printed by Thomson-Shore, Inc., Dexter, Michigan

The Conservation Foundation
1250 Twenty-fourth Street, N.W.
Washington, D.C. 20037

Library of Congress Cataloging in Publication Data
The Greenhouse effect, climate change, and U.S. forests.

1. Greenhouse effect, Atmospheric--United States. 2. Forest meteorology--United States. 3. Forest management--United States. 4. Climatic changes--United States. 5. Trees--United States--Climatic factors. 6. Plants, Effect of carbon dioxide on--United States. 7. Forest ecology--United States. 8. Greenhouse effect, Atmospheric. 9. Forest meteorology. 10. Forest management. 11. Climatic changes. 12. Trees--Climatic factors. 13. Plants, Effect of carbon dioxide on. 14. Forest ecology. I. Shands, William E. II. Hoffman, John S. (John Stephen), 1950- III. Conservation Foundation.

SD390.7.G73G73 1987 634.9'619 86-24105
ISBN 0-89164-096-7

CONTENTS

FOREWORD

Climatologists agree that the greenhouse effect and climate change is likely to affect the earth's environment in ways that will require changes in the management of our forests. *The Greenhouse Effect, Climate Change, and U.S. Forests* takes a long-term and interdisciplinary look at this situation, which may loom large over the lives and fortunes of future generations, if not our own. It represents an effort--unusual in a policy-making environment accustomed to crisis management--to get ahead of a potential problem. The contributors to this volume explore possible actions that could be taken in one area--forestry--to compensate for, and even take advantage of, an ambient atmosphere enriched with carbon dioxide (CO_2) and a changing climate.

There are, to be sure, obstacles to action. Many uncertainties are identified and discussed in the various papers included here. Moreover, change is likely to be gradual and apparent in the near term only to scientists who monitor conditions over time. Though there is evidence that increased atmospheric CO_2 and other trace gases are affecting our environment today, there is a danger that the public and even policy makers are likely to acknowledge the severity of the problem only when evidence over a long period is pieced together--for example, extraordinary frequency of drought years in an area, or the discernible loss in its traditional range of a species particularly vulnerable to environmental stress. A considerable amount of research is under way to try to remove some of the uncertainties. Yet, as a recent U.S. Department of Energy study observed, the situation requires that policy options be explored well before changes in the earth's climate become obvious. It is imperative that we look to the long-term future.

As the contributors to this volume attest, no single discipline holds all the answers. Thus, the conference that stimulated many of these papers brought together scientists working in many fields--climatologists, specialists in the composition of the atmosphere, plant physiologists, biologists, geneticists, forest economists, woodland managers, and policy analysts.

Apparently little can be done in the short term to avert the buildup of CO_2 and other gases in the atmosphere. Human-generated emissions will continue, and the gases will persist in the atmosphere for a very long time. However, the message of this report is not one of hopelessness. As Paul Kramer points out, research that seeks to deal with environmental stress like that which might occur as the greenhouse effect intensifies could help managers deal with stresses to forest ecosystems resulting from "normal" year-to-year fluctuations in climate. In other words, we will not have to wait until some indeterminate time to benefit from the results of research--new knowledge can be put to immediate use.

The greenhouse effect and climate change will continue to influence virtually every resource of the forests. Changes in forest vegetation will affect wildlife, water flows, recreation, and scenic values. The greenhouse effect will shape the economies of different regions and the amenities important to the qualities of our lives. Ultimately, it will affect where we humans live and how we use the land. This study should stimulate an appraisal of potential management responses, especially how forest managers plan for the future.

This volume discusses in one place a variety of topics of concern to forest managers and decision makers. So they can assess the potential seriousness of the greenhouse effect, it first presents the best technical information available on changes in the composition of the atmosphere, evidence of global warming, and the effectiveness of simulation models that attempt to predict future climates on global and regional scales. Next, the volume turns to challenges to forest management presented by climate change and management options. Papers specifically focus on major commercial species in the Pacific Northwest, the South, and the Upper Great Lakes, applying climatologists' predictions to what is known about the physiological responses of these species to specific environmental stresses. Finally, the potential problems are considered from the perspective of the forest products industry. Industry analysts look to the future with some confidence that current management strategies emphasizing investments in the most productive land will permit the industry to accommodate *gradual* change. However, they recognize that changes in the forest environment at the regional scale could affect returns on investments, a significant factor in a highly competitive economic environment. Much can be done to maintain management options at relatively little costs.

The idea for the conference that preceded this volume originated with the U.S. Environmental Protection Agency's Strategic Studies Staff. Earlier, that office had organized a similar effort investigating responses to sea-level rise expected as a warmer climate melts polar ice. The Conservation Foundation welcomed the opportunity to join the Society of American Foresters and the National Forest Products Association in sponsoring the conference. The participation of three very different organizations--a professional society, an industry association, and an environmental policy research organization--demonstrates that the policy and program questions related to the greenhouse effect do not fall into the province of a single organization or single element of the forestry community. Rather, the conference represented the kind of cooperative effort that should be encouraged.

Further, in this project there was a conjunction of two major concerns of The Conservation Foundation. The first is a longstanding interest in forest management and policy represented by The Conservation Foundation studies of the Eastern national forests, decision making under the Forest and Rangeland Renewable Resources Planning Act, and forest policy consensus building in the Lake States of Michigan, Minnesota, and Wisconsin. The second is a commitment to cutting-edge studies to identify emerging problems and stimulate a discussion of policy options well before they reach crisis proportions, represented by policy analysis of groundwater quality and quantity, cross-media pollution transfer, and risk analysis. As this project was proceeding, another foundation staff member wrote an article exploring the implications of the greenhouse effect and climate change on wildlife and biological diversity.

This volume, and the conference that inspired it, benefited from the expertise and efforts of John Hoffman, director of EPA's Strategic Studies Staff, who, with Conservation Foundation Senior Associate William E. Shands, co-edited this volume. Robert W. Slocum, Jr., now vice-president for forest policy and research for the American Forest Council, represented the National Forest Products Association in planning the conference, and Charles Harden, Director of Science and Education for the Society of American Foresters, represented that organization. Robin Sandenburgh of ICF Inc. also contributed to planning the conference. John Bruce Wells, of The Bruce Company, prepared several manuscripts for publication, and Bethany Brown edited the manuscript.

William K. Reilly
President
The Conservation Foundation

OVERVIEW

by William E. Shands and John B. Wells

The future global climate may be radically different from today's. We know that the global atmosphere is rapidly changing, and scientists expect changes in regional climate patterns that will influence forest distribution and productivity worldwide.

The earth's climate is not static; global temperatures have been much warmer and colder than they are now. The composition of the earth's atmosphere also has varied; the amount of carbon dioxide (CO_2) in the atmosphere, now 340 parts per million (ppm), has ranged from several thousand to as low as 200 ppm. These changes, however, have occurred on geological time scales--over thousands and millions of years. Now CO_2 and other trace gases are accumulating in the atmosphere at an unprecedented rate with effects that will be measured in decades.

Measurements show that atmospheric concentrations of CO_2 have increased 20% since 1960 and are continuing to rise. Other trace gases such as methane (CH_4), nitrous oxide (N_2O), and chlorofluorocarbons (CFCs) also are increasing. Although their sources are varied, CO_2 and these other trace gases have one common property: by trapping outgoing infrared radiation in the atmosphere, the gases act like a greenhouse to raise the earth's temperature.

Of these gases, CO_2 will have the most pervasive effect on forests. Besides contributing to the greenhouse effect, CO_2 directly affects the physiological development of trees. Paul Kramer, a plant physiologist at Duke University, believes that increased atmospheric CO_2 already is affecting forest ecosystems. The other trace gases, although not affecting plant physiology directly, may make as great a contribution to climate change. Atmospheric methane, for example, is increasing at a rate of 1% per year.

In June 1984, forest scientists, climatologists, forest industry executives, and others met in Boulder, Colorado, at a conference sponsored by the National Forest Products Association, the Society of American Foresters, The Conservation Foundation, and the U.S. Environmental Protection Agency. The objective was to exchange information on CO_2 accumulation, to assess possible changes in climate and potential effects on trees and forest ecosystems, to explore forest management options, and to identify research needs. The conference was explicitly aimed at alerting the forest products industry to impending climate change and its implications for forests.

This volume contains many of the papers presented at that conference, augmented by much new material that reflects the rapid advances in our understanding of the complex forces expected to influence future forests. However, almost every paper stresses that our understanding of the greenhouse effect still is limited. The papers, thus, recognize the uncertainty that faces the forest products industry in its responses to climate change.

OUR CHANGING ATMOSPHERE

Our heavy reliance on fossil fuels--coal, oil, and gas--began with the Industrial Revolution of the 1800s. Because these energy sources were plentiful and inexpensive, the world economy, and energy use, grew exponentially. However, the burning of fossil fuels releases carbon into the atmosphere, carbon that had been stored in the earth for hundreds of millions of years. Carbon releases from fossil fuel combustion upset the earth's natural cycle, which circulates carbon among the atmosphere, oceans, ground, and biomass. Deforestation may perturb this carbon cycle too.

Scientists agree that if atmospheric CO_2 doubles and nothing else changes the increased "radiative forcing" caused by the greenhouse effect would raise the earth's average temperature by 2.2°F (1.2°C), causing other changes. Melting ice and snow would decrease the earth's reflectivity, allowing more heat to be absorbed by the surface. Higher temperatures would increase atmospheric water vapor, itself a greenhouse gas. Scientists agree that these forces would amplify the direct radiative forcing, but they are uncertain of the exact amount. The National Academy of Sciences convened two panels to review the evidence; the panels decided that doubling the CO_2 concentration would increase the average temperature by 2.7° to 8.1°F (1.5° to 4.5°C).

GREENHOUSE GAS ACCUMULATION MAY BE INEVITABLE AND IRREVERSIBLE

Roberts (Chapter 1) points out that, for all practical purposes, the accumulation of carbon dioxide and the consequent global warming are irreversible. While the CO_2 emitted into the atmosphere by fossil fuel combustion may eventually dissolve in the oceans and become locked in carbonate rocks at the bottom of the sea, this process takes centuries. Researchers believe it reasonable to expect that CO_2 will double by the year 2050.

The global temperature increase forecast by climatologists, Roberts points out, "is huge in terms of what it will do to life on our planet." Rainfall and snowfall will be distributed differently over the earth's surface, growing seasons will change, and high-latitude glaciers and ice fields will melt. Not all regions will be affected equally. Some will become warmer and drier; others warmer and wetter. Some areas may even cool off. The continental United States will be profoundly affected, according to the doubled CO_2 scenario, with droughts becoming more frequent and severe. This portends tremendous changes in forest management.

Hoffman and Wells (Chapter 2) emphasize that fossil fuels are so intertwined with the world economy that even with the most conservative assumptions about the future--slow economic and population growth, large taxes on fossil fuels, a rapid transition to nonfossil energy sources, and a vigorous program of energy conservation--CO_2 emissions will remain high. Other trace gases are tightly linked to economic growth and their emissions, similarly, are difficult to constrain.

Hoffman and Wells, in another paper (Chapter 3), explore the contributions of other trace gases to global warming. Scientists attending an international conference in Villach, Austria, in 1985 agreed that the combined radiative effects of CO_2 and trace gases could cause a doubling of atmospheric concentrations of these gases as early as the 2030s. Thus, within 50 years there may be a global warming unprecedented in human history. While the oceans will absorb heat and thereby moderate the speed at which the warming occurs, by 2030 the average global temperature is likely to have risen from 0.9° to 2.5°F (0.5° to 1.4°C).

UNCERTAINTIES

Hoffman and Wells emphasize that projections of future trace gas emissions and concentrations are fraught with uncertainty. Some emissions, particularly methane, are from natural sources that we cannot yet quantify. Other emissions are generated by humans and the amount of emissions will depend on economic growth, population growth, technological progress, and other societal factors. And we do not have a good understanding of natural biological, geological, and chemical cycles that influence the concentration of CO_2 and other gases in the atmosphere.

Climate change analyses are themselves uncertain. Hansen and his colleagues (Chapter 4) discuss the climatologists' state-of-the-art tool, the general circulation model (GCM). These models are mathematical representations of the broad-scale atmospheric processes that control regional climates. GCMs are designed to predict climate variables at all locations in the world and to show changes in hydrological processes, such as precipitation, evaporation, and runoff. This is a daunting task. Current GCMs can mimic many current climatic regimes in the world, but they cannot successfully reproduce current conditions in many locations. There is not enough information on how clouds affect global climate, for example; nor do scientists have the capability at present to model ocean response to temperature change.

Most uncertain are regional climate change projections. Rind (Chapter 5) notes that historically North American regional climates have varied significantly. Even during the Ice Age, when glaciers covered much of the North American continent, there was significant regional variation, as shown by geological studies of fluctuations in lake levels in the southwestern United States. Comparisons of warm and cold periods in the 20th century show considerable variations in precipitation in different regions of the United States.

Without accurate regional climate projections, the effects of climate change on specific tree species in specific regions cannot be analyzed. Rind cautions that "different models vary in their prediction of what changes will occur from place to place." Until the GCMs can be improved, Gibbs and Hoffman (Chapter 6) suggest developing hypothetical scenarios that approximate expected climate change to assess impacts. They describe a methodology for developing alternative, but internally consistent, scenarios.

EFFECTS ON FORESTS

How rising CO_2 and potential climate change might affect forest management is discussed in an excerpt from a report prepared by Sandenburgh and colleagues (Chapter 7) for the U.S. Environmental Protection Agency. The authors' main point is that whether forest productivity decreases or increases will depend on how forest managers respond if atmospheric CO_2 changes natural growing conditions. Managers will have to change their planning and decision-making processes, they suggest.

Just how climate change might affect important commercial species in two of the nation's major timber-growing regions is explored in papers by Leverenz and Lev (Chapter 8) and by Miller and colleagues (Chapter 9). Leverenz and Lev project distribution under a scenario of doubled atmospheric CO_2 for six western U.S. tree species--ponderosa pine, Douglas-fir, western hemlock, western larch, lodgepole pine, and Engleman spruce. Their analysis is based on observed physiological changes in plants induced by accumulating CO_2 combined with rising temperature and water stress. These projections are sensitive to assumed responses to the balance between precipitation and evaporation on a site and the assumed chilling requirements of different species.

In brief, Leverenz and Lev project that there will be changes in the range of a species, as well as increases or decreases in importance over sections of its range.

Miller and his colleagues found that loblolly pine would be affected significantly in terms of both range and productivity. Its range is likely to move north and northeast, into eastern Pennsylvania and New Jersey, and shrink in the west, withdrawing into central Arkansas. The total range of the species is likely to increase from the present 344,000 square miles (139,271 hectares) to 385,000 square miles (155,870 hectares). Despite the increase, they project a net loss of commercial loblolly lands because a large proportion of the new range will be in highland sites less accessible and less productive than present loblolly sites in the Coastal Plain.

Solomon and West (Chapter 10) describe a computer model that simulates some interactions among tree species--specifically, birth, death, and growth characteristics--to analyze the effect of environmental stresses on productivity. Applying the model simulations to sites in three regions--central Ontario, the western upper peninsula of Michigan, and middle-eastern Tennessee--they postulate changes in forest composition and productivity that would lead companies to retain, sell, or acquire land in those areas. The complex results should intrigue forest industry decisionmakers. Solomon and West are quick to point out weaknesses in the simulation approach but contend that it is potentially a powerful analytical tool for forest products companies.

Kramer and Sionit (Chapter 11) investigated offsetting effects of increased CO_2. Based on their short-term experiments at Duke University and the work of others, Kramer and Sionit report that increasing the concentration of atmospheric CO_2 increases stem diameter, height, and dry weight of seedling of several tree species; and branching, leaf area, and leaf thickness in some, but not all, species. "It seems probable," they conclude, "that if other environmental factors are not severely limiting, the increasing concentration of atmospheric CO_2 will significantly increase tree growth, but the magnitude of the effects will vary among species and with the severity of other environmental stresses."

INDUSTRY RESPONSES

Obviously, all this raises potentially troublesome long-range problems for the forest products industry (and corresponding problems for managers of public forestland). Sandenburgh and colleagues (Chapter 12) describe the environment for industry decisionmaking and suggest possible industry strategies to cope with rising CO_2 and climate change. They argue that long-range planning is economically sound, and describe a framework for analysis of strategic and operational decisions, ranging from land acquisition, to mill siting, to scheduling of timber harvests.

Papers by and Rose and colleagues (Chapter 13) and Woodman (Chapter 14) look at the situation from the position of an industry preoccupied with volatile, short-term economic problems. Rose and his colleagues agree that "it is not too early, however, to organize available information to identify major gaps in knowledge and develop a conceptual framework for analysis" despite inadequate data. Strategies should emphasize flexibility in response to change.

Using the limited information available, Rose and colleagues focus on possible impacts, under the doubled CO_2 climate scenario, on the aspen-based forest industry of the Great Lake States and the loblolly pine forest industry of the southeastern United States. In the Lake States, shortfalls of aspen can be covered by other presently underutilized hardwoods and aspen stand residues. In the Southeast, however, climate change could have a major impact on supply, since shortfalls are currently predicted, irrespective of climate change, in both softwoods and hardwoods by the year 2030. However, companies now are reevaluating their land ownerships and management programs, and climate change scenarios are not likely to affect current strategies to deal with that expected supply shortfall.

Woodman too foresees no major changes in management of Douglas-fir and western hemlock on the coast ranges and west slope of the Cascade Mountains in Washington and Oregon, as a result of CO_2-induced climate changes. He notes that the transition from harvesting of old-growth natural forests to the tending of human-established stands will be complete in the not-too-distant future. The trend, now under way, toward prescribing management practices on the basis of site productivity--with intensified management of stands on highly productive sites--should allow forest managers to adjust to gradual change. Nonetheless, Woodman does expect management problems in forest regeneration at an affordable cost in areas where water stress occurs.

Woodman concludes that forest management will change, but more for economic than biological reasons. In a volatile economic climate, a company's first consideration is survival. He urges accelerated development of regional or local climate prediction models and long-term research on the costs and benefits of management practices in different local climates and higher levels of CO_2.

Tree improvement, specifically the breeding of trees that can survive, and perhaps flourish, under climate conditions of elevated CO_2 is one response to changing conditions. But, because of the long time required for forest trees to attain sexual and economic maturity, the geneticist cannot react quickly enough to avert catastrophe. As Kellison and Weir (Chapter 15) point out, current greenhouse tests on the effects of elevated CO_2 are conducted under conditions of ample light, moisture, and nutrients--conditions that might not be present if climates change as some climatologists project. However, the "genetic diversity within a wide-ranging tree species, such as loblolly pine, is sufficiently broad to allow for directed management in new environments that fall short of catastrophic levels."

RESEARCH NEEDS

In the final paper, Lee and Kramer (Chapter 16) argue that by anticipating changes "the forestry community can avoid surprises, identify alternatives, and seek to temper ad hoc political decisions with technological information." Potential problems of increased CO_2 and climate change have low priority among forest investors, managers, and public administrators, they point out. Many factors contribute to this lack of urgency: the problem is global and overarches the responsibilities of many federal agencies with no one agency clearly in charge; scientists in numerous disciplines are working on various aspects of the problem and do not communicate easily; and investment pay offs are uncertain.

Lee and Kramer argue that, with or without CO_2-induced climate change, research on problems of CO_2 accumulation and climate change will help answer a number of contemporary questions about how trees respond to environmental stress and how adverse effects can be mitigated. Most gaps in knowledge, they contend, are amenable to conventional research methods. They then outline a strategic approach to research.

First, there is a need to recognize that increasing atmospheric CO_2 is affecting tree growth now; forest products industries should not consider it a distant problem to be addressed some time in the future. Second, the direct effects of CO_2 build-up should be separated, for research purposes, from indirect, climate change effects; while there are uncertainties over the timing, magnitude, and regional effects of global warming, direct fertilizing effects of increased atmospheric CO_2 are being documented now. Third, since relatively short-term variations in temperature and precipitation will have more influence on tree growth than gradual greenhouse-induced climate change, plant physiologists and forest scientists should concentrate on understanding how current weather extremes affect the physiological processes that control tree growth.

Lee and Kramer also recommend increased research on how weather affects the forest environment; more basic research on tree growth processes and how these are affected by increased CO_2; new approaches to securing funds for forestry research from diverse agencies by tailoring proposals to each agency's mission; and greater communication among forest scientists, crop scientists, and climatologists. The fundamental question, they conclude, is not whether scientists will be able to provide conclusive proof of future risk, but whether society can and will take the action necessary today to address changes or problems that will confront it some time in the future.

A final note: Although it focuses on the timber industry's response to increased CO_2 and a changing climate, this volume should be of interest to a wide range of forest interests. CO_2-induced changes in forest environments will affect forest recreation, wildlife, and aesthetic values. Research and discussion of the effects of climate change on other forest values deserve high priority.

TIME TO PREPARE FOR GLOBAL CLIMATIC CHANGE

by Walter Orr Roberts

ABSTRACT

Warming of the earth from rising levels of carbon dioxide (CO_2) and other "greenhouse gases" in the atmosphere, caused by the burning of fossil fuels and the release of other trace gases, will affect many aspects of our lives. Though climatic change is believed inevitable by many climatologists, there are many uncertainties in its predicted magnitude and timing, largely because of imperfections in current global climate models. Climatologists estimate, however, that the ambient global temperature is likely to increase 3 to 4°C (5 to 7°F). If this occurs, agricultural and forested areas in the United States and Russia will become warmer and drier than they are now, and the north polar ice cap will melt. There will be winners and losers. While one can conjecture about continental effects, there will be extreme changes at the regional level. Weather variability may decrease, but will still be highly diverse. Somber as the picture is, human societies can accommodate to changes in climate of the magnitude expected. Strategies include planting of trees on lands marginally adequate for irrigation, greenhouse agriculture, and genetic engineering of plants which can survive in water with high salt content. These strategies will pay off even if the climate and weather remain as they are now--fickle and violent systems that strongly influence the course of human affairs.

There is mounting scientific evidence that significant warming of the earth is in prospect in the last years of the 20th century and may bring a new and different climatic regime without parallel in the recent history of the planet. This warming will affect agriculture, forestry, livestock, water, transportation, energy use, building construction, and many other aspects of human life.

Scientists agree, by and large, on these facts:

- Carbon dioxide (CO_2) and other trace gases are accumulating in the atmosphere.
- This accumulation will cause a gradual warming of the average global temperature of something between 1.5 and 4.5°C

(3 and 8°F) by sometime in the next century.

- With this gradual warming will come a change in patterns of wind, rain, snowfall, and other elements of climate.
- There will be considerable regional climatic variability, with some regions becoming warmer and drier, others warmer and wetter, and some perhaps even cooler because of the influence of oceans.

There are, however, many uncertainties. Estimates of the magnitude of future warming and its pace depend on simulated models of the global climate. These are far from perfect and are not able to incorporate all significant variables in their calculations, particularly the response of oceans to changes in surface temperatures and the effect of clouds on the earth's reflectivity. Further, they are not able to project temperature and climatic change reliably at the regional level.

Even so, many climatologists believe, from modeling research, that global warming is inevitable. Although a warming of 3 to 4°C (5 to 7°F) may not sound like much, it is huge in terms of what it will do to life on this planet. The central parts of major continents in midlatitudes (the corn and wheat belts of the United States; the wheat-growing regions of the USSR) are likely to be appreciably warmer with more severe hot spells and lower soil moisture; the tropical rain belts may expand, bringing benefits to the sub-Sahel and perhaps to parts of South America; the north polar ice cap may melt completely in summer sometime in the next century, although melting of the Antarctic ice will take far longer.

GROWING SCIENTIFIC INTEREST

The buildup of atmospheric CO_2 and its implications for the earth's climate has been recognized for decades. While scientists have been measuring increases in CO_2, climatologists have been studying the trend of global temperature. In the late 1970s there was a flurry of reports that attempted to synthesize available information (e.g., *Energy and Climate*, Geophysics Study Commission, National Research Council, 1977; *Carbon Dioxide and Climate: A Scientific Assessment*, 1977).

In 1983 the National Research Council of the National Academy of Sciences issued *Changing Climate*, which, in the words of the chairman of the scientific committee that produced the document, was an effort to "achieve a comprehensive and internally consistent assessment" of the problem. *Changing Climate* reverses and synthesizes much of the available information on fossil fuel use, accumulation of CO_2 and other trace gases in the atmosphere, climate effects and their detection, effects on agriculture, water, and sea level, and social implications. The Department of Energy also has sponsored considerable research on the subject. And, as this volume indicates, a number of scientists are at work on different aspects of the problem.

TEMPERATURE RISE IRREVERSIBLE

The main source of carbon dioxide buildup is thought to be fossil fuel burning, although progressive deforestation also may contribute appreciably by lowering the absorption of atmospheric CO_2. Today the atmosphere contains about 340 parts per million (ppm) of carbon dioxide, up 20 to 25% since the start of the industrial revolution, around 1860. Since 1860, energy usage has been growing steadily at over 4% per year, although after 1973, when oil prices skyrocketed, the rate of growth dropped to perhaps half that amount.

Even with the large increases in energy costs, there is no clear prospect of any long-term downturn in energy use, though conservation measures adopted in industrialized countries have helped lower energy-use growth curves. Moreover, recent declines in oil prices may accelerate usage. The economic improvements now emerging in a number of developing countries likely will further stimulate global consumption of fossil fuel.

For all practical human purposes, the carbon dioxide warming is *irreversible*. People are not going to stop burning fossil fuels. And while the carbon dioxide that enters the atmosphere from human activity eventually dissolves in the oceans and then precipitates out as carbonate rocks in the sea bottom, the process takes many hundreds of years. If the carbon dioxide burden in the air doubles by 2050, it will probably take until the year 3000 for the natural "cleanup" by the sea to restore the present level, even if fossil fuel usage drops to the pre-industrial level--a long time in terms of human history.

The oceans do slow down any climate warming or cooling, because the great heat content of the oceans makes them respond slowly to an atmospheric temperature change. Thus the oceans act like a flywheel smoothing out the changes of speed in an engine. If the greenhouse effect were suddenly to increase, the ocean's inertia probably would delay the response of global temperatures to the new equilibrium by 10 to 20 years.

Oil and natural gas are superb fossil fuels for today's needs. Both are economical and can be burned relatively cleanly if attention is given to containing airborne pollutants. But reserves of these fuels are disappearing rapidly, despite recent discoveries of additional resources and despite all the newly adopted conservation measures.

Thus, coal seems to be the most likely bridge to an energy future that is 50 or 100 years distant, in which solar, wind, nuclear, biomass, and other renewable sources power the bulk of society's machinery. There are huge world reserves of mineable coal. The joker is that coal, though abundant, produces considerably more carbon dioxide than oil or gas to generate the same amount of energy. Thus it is only a matter of time, in the view of most experts, before the carbon dioxide level of the atmosphere exceeds 600 ppm.

The date for this event is usually put at some time between about 2025 and 2075, depending on how effectively high costs and conservation keep down energy consumption and on how fast the developing world industrializes. The later dates are expected by those who do not foresee a continued large growth in fossil fuel consumption over the next half-century. A few observers even believe that the carbon dioxide problem can be fully averted by fuel conservation and use of renewable energy sources that are, at the same time, less costly. Unfortunately, this scenario is a minority view.

Moreover, the other greenhouse gases will almost certainly augment the CO_2 warming. These gases, some dozen or more of them, may become as important as CO_2 by the middle of the next century. Thus, they will further elevate the global warming. Among the gases are methane and the chlorofluorocarbons (Freons) used in industry. All of these gases are increasing in atmospheric concentration, and in aggregate they are expected soon to equal CO_2 as a cause of warming.

WHAT IT WILL MEAN FOR THE EARTH'S WEATHER

Almost everything known about the climatic effects of the current CO_2 situation rests on the theory of the "greenhouse effect." There is no historical or even clear prehistorical precedent on which to rely. But the theory has a solid base, and calculations by many different groups lead to about the same expectations so far as the global temperature average is concerned. As mentioned, the doubling will produce, it is believed, approximately 3°C (5.5°F) warmer global temperatures, with the strongest warmings at the highest latitudes and perhaps only 1 to 2°C (2 to 3.5°F) in the equatorial zone. The uncertainty in this estimate means that the warming might be as little as 1.5°C (3°F) or as great as 4.5°C

(8°F), within the limit of existing theoretical calculations. The effect of the other greenhouse gases is similar.

What does a level of atmospheric CO_2 twice that of the pre-industrial revolution period actually imply? There are some big uncertainties about how the higher levels of carbon dioxide and the other gases will affect the climate and weather of different geographical regions. After studying volumes of data and various analyses of the carbon dioxide impact, one conclusion I reach is that the effects in many areas of the earth will be large, measured in terms of what human societies will have to do to compensate. Agriculture will be most strongly affected. The temperature rise itself will be a far smaller problem than the changed rain and snow and wind patterns produced by the warming and the occurrence of hot spells, which are likely to be more severe than now. There will be winners and losers. Some countries will benefit, others will suffer.

It is probable that the overall change will be made up of many regional changes of more extreme magnitudes affecting small areas based on local mountain, ocean, forest, or ice environments. Local changes will be apparent in temperature, cloudiness, overall precipitation, humidity, soil moisture, wind direction and speed, storm frequency, and length of growing season--factors that make up the overall climate at a given time and place. Some places will be warmer and some colder, some wetter and some drier--and averaged they will equate to the slightly warmer total global temperature. Today's mathematical models cannot predict local details with any certainty. However, a strong research push is under way because people everywhere are conscious of the practical importance of having sound, quantitative, regional-scale answers as soon as possible.

Two principal large-scale climatic effects at least seem likely. First, the far greater heating of the poles compared to the equator will weaken global wind systems, because they are driven by the temperature contrast between the cold poles and the warm equator. The wind shift probably will expand the tropical moist zones. The dry belts now predominantly at 25° to 30° latitude in the Northern Hemisphere also can be expected to move northward. The result will be less precipitation in continental regions from 30° to 40° latitude as the weather warms, further increasing evaporation and worsening the water stress on trees and plants. This is very bad news for the corn and wheat belts of the United States.

Second, weather variability may decrease, though this is speculative. The winters may become weaker in circulation and, thus, more like summers, which are characterized by weaker circulation as well as by hotter temperatures. At first sight, this seems like a good thing. However, many rain-fed crop regions depend on a relatively frequent cycle of soaking rains and drying sunshine. If the sunny spells or the rains, either one, show less variation, droughts or excess rains may be prolonged.

Despite all of this, the weather will still be highly diverse. The overall warming will be the product of some regions that are cooler (perhaps from increased cloudiness near the seacoasts) and others that are more than correspondingly warmer. Moreover, there is no likelihood that the weather will become placid, without blizzards, tornadoes, floods, and hurricanes.

The wheat and corn belts of the United States are likely to suffer from increased drought frequency and severity, and there is probably no good prospect for greatly expanded irrigation, especially in the wheat belt. The "Great American Desert" depicted east of the Rocky Mountains in early geographic maps may well be real by 2040 as summers heat up and year-round weather dries up.

If the greenhouse warming produces the effect considered most probable, the Dust Bowl of the 1930s may seem like child's play compared with the "dust bowl" of the 2040s. Natural rainfall could decline by as much as 40%, and the summers will be hotter, increasing the evaporation of soil moisture. The soils will desiccate, and the winds will lift them to the skies.

By then the mineable waters of the Ogallala aquifer, which now provides much of the water for agriculture in the Great Plains, will be gone. Irrigation water of good quality will escalate in price from perhaps $15 per acre-foot to more than $200 (in 1984 dollars). This will lift pumped groundwater out of the realm of practicality for agriculture. And what then will happen to the Great Plains' farmlands? They may revert to desert. A great industrialized nation like the United States perhaps will see desertification of the kind usually associated with poor, technologically backward, developing nations.

The Colorado River Basin is another problem area. The waters of the Colorado River Basin are critically important to agriculture, industry, residential use, and recreation. And these waters are overallocated, even for the river's present average flow. It would not be surprising to see, by 2020, the water flow from the Upper Colorado decline to as little as 60% of its average flow today, with disastrous consequences.

Global warming is expected to be most severe in higher latitudes, where the temperatures may rise 10°C (18°F) rather than 3 or 4°C (5.5 or 7.0°F). The north polar sea ice could melt, leaving an ice-free summer Arctic and far safer high-latitude shipping lanes. East Africa and the Sahara may be wetter, a boon to food production on the African continent.

Some scientists have warned of a surge in heights of oceans if the West Antarctic ice sheet slides into the sea after the warming melts the ice at the sheet's periphery. This cannot be ruled out, and the consequent 5-m ocean rise would indeed cause catastrophe in many regions. However, the likelihood of its happening in the next few hundred years is remote, and thus it is a far less overall threat to humanity than the prospect of increased aridity in many of the world's best food-producing regions.

ACCOMMODATING TO CLIMATE CHANGE

This is a somber picture of the world's future. However, strategies exist and others can be developed to accommodate successfully to climate change, even of the magnitude expected. In fact, a host of sound policies and strategies are available to make human societies more resilient to the climate fluctuations they already face from year to year, even without the bugaboo of increased carbon dioxide and other greenhouse gases. Climate and weather go through major changes on every space and time scale. One winter is cold and snowy, another arid and warm. One region has a drought, while another thousands of miles away has abnormally heavy rains.

Such anomalies, historically, are the principal cause of crop failures and attendant famines. Therefore, improving the weather resistance of agriculture, forestry, and other human enterprises makes immediate common sense. Moreover, such strategies make good economic, political, and social sense even if no carbon dioxide warming comes. Any honest scientist today will admit to being unsure about any specific greenhouse effects. Quite reasonably, most industry and government planners will adopt a wait-and-see attitude about spending money now to avert an uncertain hazard rather far in the future.

On the other hand, protective steps will pay off their costs, albeit at a slower rate, if climate changes do not materialize. Moreover, measures can be taken that will maximize the benefits to some societies if climate changes do happen. And these steps, too, can pay their way if the climate and weather remain just as they are now, fickle and violent systems that strongly influence the course of human affairs. The time has come to embrace existing opportunities to build more climate-resilient human societies.

Many working agricultural lands have marginally adequate water for irrigation, but excellent soil. Many of these can be forested, producing substantial biomass. The trees, as they gain size, would provide a microclimate for other forms of life, both plant and animal.

Wood products are always in demand. Excellent liquid fuel can be made from wood and wood wastes, as can a wide variety of valuable chemical products. For decades, automobiles have been made to run cleanly and efficiently on alcohol, methane, and charcoal, all of which are derivable from wood and agricultural wastes. Wood-derived fuels hold a vast new potential for industry to explore. They have environmental benefits and probably are not far from being economically competitive, especially in places enforcing tight environmental standards.

To be sure, agricultural yields must be increased on the remaining good soil, and factory production of foods by fermentation and other novel industrial processes will have to assume greater importance. Food needs will mount even faster than population, unless the world is to remain content to let many millions starve. But the potential for providing more and better foods is great. Irrigation is but one strategy, when and where natural resources permit. Another is closed-environment agriculture: plastic greenhouses to conserve water, reduce pesticide needs, and provide additional crops per year. Many greenhouse crops can be grown with 10% of the water needed for field crops. Costs are high but rapidly are becoming competitive in well-planned systems. The arid Southwest of the United States will see huge increases in this form of food production.

Still another technique is growing halophytes--trees and other plants that thrive in water too salty for ordinary crops. And there are thousands of miles of arid, sunny seacoast where grains, forage plants, and trees can be grown with full-strength sea water for irrigation. Moreover, the large portions of conventionally irrigated croplands with high saline contents, caused by the shortage of good fresh water to flush out the salts that result from evaporation, could perhaps support halophytes. Many of these trees and plants produce yields competitive with corn or wheat in conventional field culture, and the grains are of high protein and other nutritive values. Genetic engineering can probably be counted on to bring a new "green revolution" to the halophytes, but the first gun has yet to be heard.

And many other policies and strategies can help. Conservation of energy could slow down the carbon dioxide buildup by lowering fossil fuel consumption. Soil conservation is another. Eroded soil also releases carbon dioxide, and carefully managed soil is more productive and permanent.

Growing new forests, enhancing the productivity of existing forests, and ending deforestation are still other worthwhile strategies. Controlled forestry can, as many private companies and government programs demonstrate, keep forestland in a state of constant renewal.

So there are good, solid reasons to believe that the planet's inhabitants can live comfortably in a warmer, drier world. However, the political challenge posed by the prospect of a new and warmer climate is just the kind that nation states are least equipped to meet. It is a slowly approaching, almost invisible issue of uncertain timing and magnitude. There is no spectacular hazard crying for attention, unlike acid rain. No technological "fix" is in the wings. The phenomenon is global, rather than national. Strategies to avert its dangers are expensive and have lead times measured in decades rather than political terms of office. Yet there are strategies available to conquer the tyranny of drought, famine, and flood and to overcome the future stresses which are likely from a manmade and irreversible change of climate.

REFERENCES

Geophysics Study Committee. 1977. *Energy and Climate.* National Academy Press, Washington, D.C.

National Research Council. 1979. *Carbon Dioxide and Climate: A Scientific Assessment.* National Academy Press, Washington, D.C.

FORESTS: PAST AND PROJECTED CHANGES IN GREENHOUSE GASES

by John S. Hoffman and John Bruce Wells

ABSTRACT

Recent atmospheric measurements demonstrate that the atmospheric concentrations of many "greenhouse gases" are increasing. Future concentrations will depend on the growth of emission sources, current emission factors for each source, possible changes in emission factors, biogeochemical cycles that control the fate of emissions, and possible changes in these biogeochemical cycles. Because of the uncertainties inherent in predicting the future, researchers have developed alternative scenarios for the growth of these gases.

Carbon dioxide (CO_2) concentrations have risen 25% since the onset of the Industrial Revolution. Future concentrations will depend on fossil fuel use and how this carbon is cycled in global ecosystems and the oceans. CO_2 concentrations will probably double by the mid-21st century.

Chlorofluorocarbons (CFCs) are synthesized chemical compounds used in a variety of goods and processes. Unknown before the 1930s, their concentrations have doubled in the last 10 years. Future emissions are intimately linked to economic growth. Because CFCs build up in the atmosphere, concentrations will continue to rise even if emissions are held constant.

Methane concentrations have roughly doubled in the last 150 to 200 years and are currently increasing at about 1% per year. While rough estimates of emission sources are available, the uncertainties about future growth are large.

Nitrous oxide (N_2O) concentrations arre increasing at about 0.2% per year. The major sources of N_2O are fossil fuel combustion and fertilizer use. As with CFCs, the atmospheric lifetime of N_2O is long, and concentrations will grow even if emissions do not.

Researchers have identified a score of other radiatively active trace gases. Taken together, CO_2, CFCs, methane, N_2O, and these other gases have the potential to significantly alter the earth's radiative balance.

Recent atmospheric measurements demonstrate that the global atmosphere is not static, but is a dynamic system sensitive to man's influence. Measurements show that the concentrations of many "greenhouse gases" (gases with the potential to raise the earth's surface temperature by reducing the emission of heat to space) are increasing. The increased concentrations of trace gases in the 1970s alone may eventually contribute to an increase in the average global temperature of 0.2 to 0.3°C (0.4 to 0.5°F) (Lacis et al., 1981). While much of this increase is due to carbon dioxide (CO_2), other trace gases play an equal role.

This chapter reviews the trace gases likely to be important contributors to future global warming and associated climate change. Each gas is considered in the order of its potential effect on climate, based both on its effectiveness in absorbing radiation and on its future concentrations. While absorption is scientifically determinable, future concentrations will always be subject to debate. Predicting such changes requires an understanding of both the sources and the fates of emissions, including the biological, chemical, and geological processes that transport, deposit, or transform chemical species, the so-called biogeochemical cycles.

CARBON DIOXIDE (CO_2)

Measured Increases in CO_2 Concentrations

Tyndall noted the radiative properties of water vapor as early as the 1860s (Tyndall, 1861). Since his early work, CO_2 is the greenhouse gas that has received the most attention and is still thought to be the most important radiatively active gas. Consistent and accurate measurements of its concentrations began in 1958 at Mauna Loa Observatory in Hawaii (C. D. Keeling, personal communication, 1985). The data show an increase of approximately 10% (Figure 1).

While direct measurements of CO_2 concentrations were made as early as 1872, they are subject to error (Callendar, 1958). Indirect methods such as measurements of carbon dioxide trapped in ice cores (Delmas et al., 1980; Oeschger et al., 1982; Neftel et al., 1982) and analysis of isotopic carbon ratios in tree rings (Freyer, 1978; Peng et al., 1983) show that CO_2 concentrations before 1850 were constant at approximately 270 ± 10 parts per million (ppm), 25% lower than today.

Figure 1

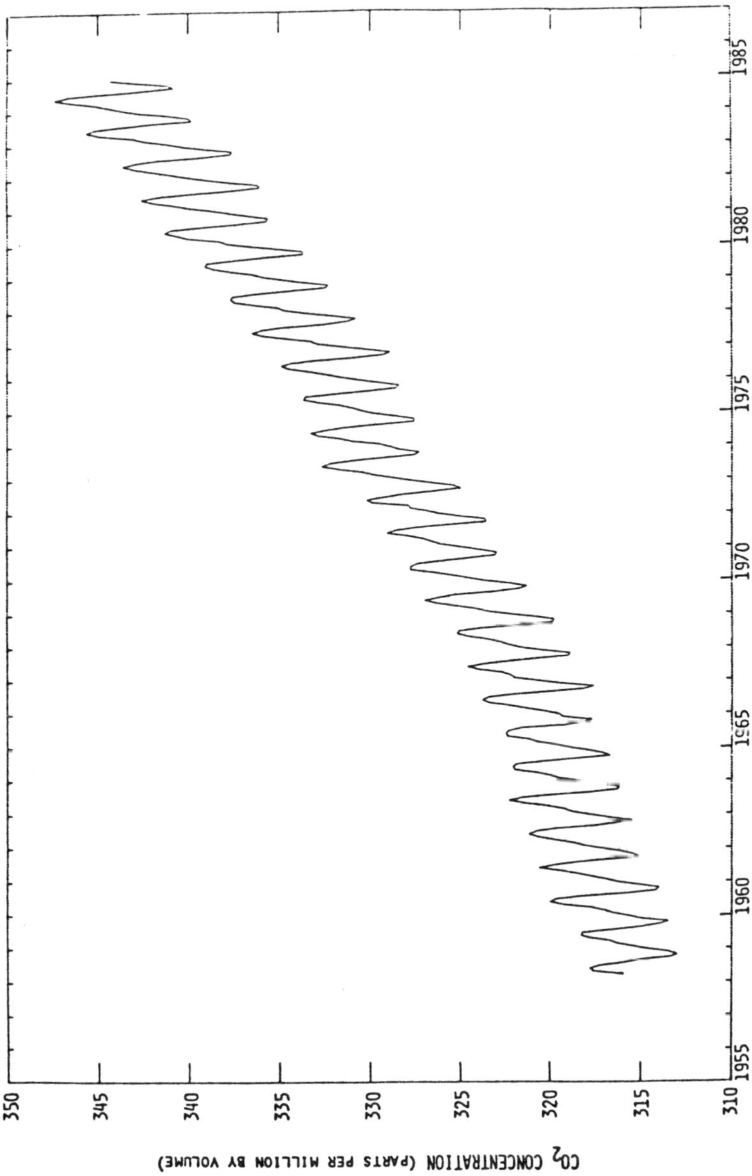

Since 1958, CO_2 concentrations have risen approximately 10% (Keeling, personal communication).

Historical CO_2 Emissions

Before 1850, the natural losses of CO_2 presumably balanced the natural emissions, leading only to a seasonal cycle still visible each year. After 1850 and the onset of the Industrial Revolution, the heavy use of coal and other fossil fuels (and possibly massive deforestation) led to a large increase in carbon dioxide emissions. Rotty and Marland (1984) have analyzed historical records of fossil fuel use and reconstructed the emissions profile (Figure 2). Such activities as forest burning and deforestation also emit CO_2, although recent estimates of this possible source have declined greatly in recent years (Brown and Lugo, 1981; Houghton et al., 1983).

The Carbon Cycle: Biogeochemical Factors

Direct emissions of CO_2 from consumption of fossil fuels are not the only source of atmospheric CO_2. The biochemical processes known as the carbon cycle include the terrestial biosphere, i.e., plants and trees which trap CO_2 in photosynthesis and release it in decomposition and burning, and the oceans, which both absorb and release CO_2. Figure 3 illustrates the relative importance and interactions of these components. Currently it is believed that, over time, the fluxes between compartments can change, for example as the oceans become saturated with CO_2, decreasing their absorption, or as increased photosynthesis increases stored carbon in biomass. The percentage of worldwide carbon dioxide which remains in the atmosphere (the "fraction airborne") is between 40 and 60% (Clarke, 1982). At least some researchers believe this amount will increase (Emanuel et al., 1981).

Projections of Future CO_2 Emissions

Future concentrations of CO_2 will depend on future CO_2 emissions and the future operation of the carbon cycle and its impact on the fraction airborne. Consequently all predictions of future CO_2 levels are subject to uncertainty. To project future CO_2 emissions, various researchers have used mathematical models which attempt to represent, in the aggregate, the socioeconomic factors which contribute to energy use, usually connecting world gross national product and population with energy use and economic efficiency. By making different assumptions about future populations and economic growth, the modelers generate projections of future energy use. They then apply average carbon content ratios to

Figure 2

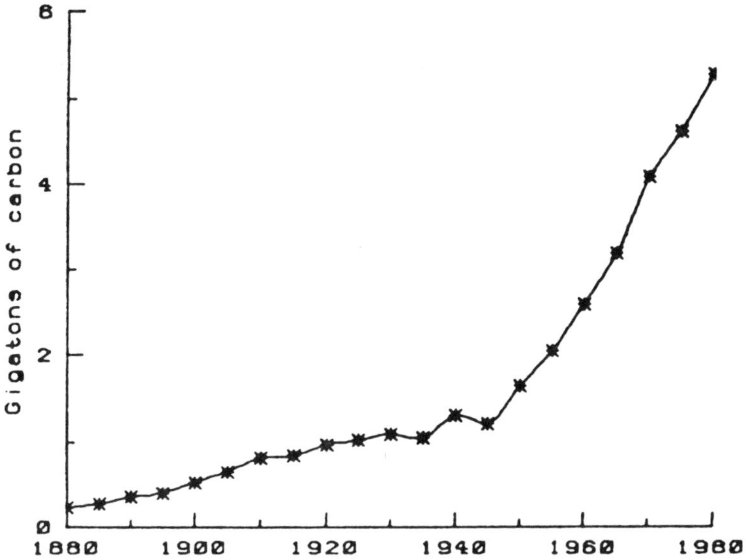

The Industrial Revolution and its heavy reliance on coal led to a large increase in CO_2 emissions (Rotty and Marland, 1984).

Figure 3

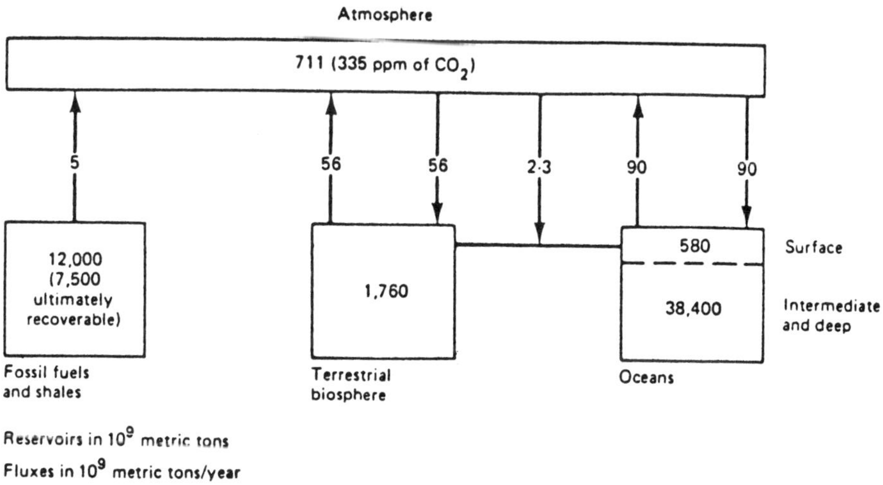

Taken together, these stocks and flows constitute the carbon cycle. Carbon distributed through this cycle plays an important role in the ecological balance of the atmosphere, biosphere, and oceans (Council on Environmental Quality, 1981).

compute the associated CO_2 emissions.

Perhaps the most widely accepted set of projections was developed for the National Academy of Sciences by Nordhaus and Yohe (1983). In their most likely scenario, CO_2 emissions rise from 5 gigatons per year in 1975 to nearly 20 by the year 2100. Other researchers with published projections of CO_2 emissions include Seidel and Keyes (1983) of the U.S. Environmental Protection Agency and Edmonds et al. (1984) of the Institute for Energy Analysis. Figure 4 shows the low, medium, and high projections of each group.

Projections of Future CO_2 Concentrations

As carbon dioxide emissions increase, researchers expect that the airborne fraction may increase as the major sink for CO_2, the upper ocean, becomes saturated (Emanuel et al., 1981) or deep-water formation in the North Atlantic, the principal mechanism causing the exchange of CO_2 between the deep ocean and surface layers and atmosphere, slows (Broecker and Takahashi, 1983). Thus, both rising CO_2 emissions and possibly an increasing airborne fraction should ensure future increases in CO_2 concentrations. A widely adopted measure of concentration projections is the "doubling time," the year in which CO_2 concentrations reach twice the preindustrial level. While the range in emissions projections looks wide, the range in doubling time, shown on each curve in Figure 4, is smaller. For the medium cases reported by each group, the doubling time occurs near 2060. To put the doubling in perspective, one should recall the main conclusion of the latest National Academy of Sciences panel: a doubling of CO_2 would ultimately cause an average warming of 1.5 to 4.5°C (2.7 to 8.1°F).

CHLOROFLUOROCARBONS (CFC 11 AND CFC 12)

Measured Increases in CFC 11 and CFC 12 Concentrations

CFC 11 ($CFCl_3$) and CFC 12 (CF_2CL_2) are synthesized chemical compounds used as aerosol propellants, solvents, blowing agents for insulating foams, and in a variety of other industrial processes. Recent atmospheric measurements show that their concentrations have been increasing approximately 5% per year since 1978 (National Aeronautics and Space Administration, 1986) (Figure 5).

Figure 4

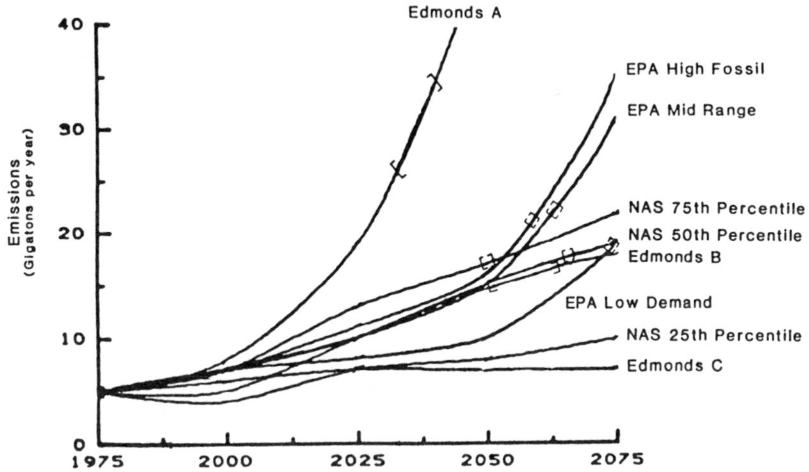

Brackets show approximate timing of doubled CO_2 concentrations

The medium cases of these researchers show that a doubling of atmospheric CO_2 concentrations will occur near the year 2060 (Nordhaus and Yohe, 1983; Seidel and Keyes, 1983; Edmonds et al., 1984).

Figure 5

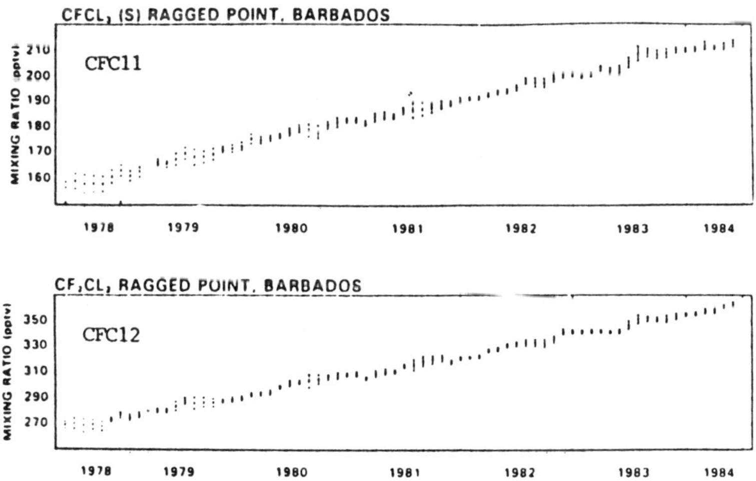

Barbados is a clear site representative of the global atmospheric concentrations of CFC11 and CFC12, which have risen approximately 5% per year since 1978 (National Aeronautics and Space Administration, 1986).

Historical Use and Emissions of CFC 11 and CFC 12

CFC 11 and CFC 12 have enjoyed rapid growth since their discovery in 1931 (Figure 6). Researchers have clearly demonstrated that the use and emissions of CFC 11 and CFC 12 are intimately linked with the world economy. Perhaps Figure 7, with its simple presentation of U.S. industrial output from 1967 to 1983 and nonaerosol CFC 11 and CFC 12 use in the same years, most strongly demonstrates this relation. Quinn et al. (1985), in a study commissioned by the U.S. Environmental Protection Agency, found that, on average, a 1% change in gross domestic product in developed nations was followed by a 2 to 2.5% change in nonaerosol use. Figure 8 shows the relationship between gross domestic product in developed countries and nonaerosol CFC use over the last 20 years.

The distinction between aerosol and nonaerosol use is important because many Western nations, fearing that CFCs may endanger the stratospheric ozone layer, curtailed their use of CFCs in aerosols. In 1978 the United States banned their use in nonessential aerosols, and since that time the nations of the European Community have undertaken voluntary measures which have reduced their aerosol use by a third. As shown in Figure 6, these one-time reductions counteracted the continuing growth of CFCs in nonaerosol uses to produce a constant overall emissions profile during the mid-1970s to early 1980s. CFC use in nonaerosol applications, however, continues to grow, and overall emissions are beginning to increase. The use of CFC 11 and CFC 12 in the United States has grown at a 15% rate for the last two years (U.S. International Trade Commission, 1985).

CFCs are well suited to many industrial processes because of their stability; they are nontoxic, nonflammable, and virtually indestructible. It is this stability which heightens their effect on climate. After they are emitted, CFCs slowly drift upward in the atmosphere, where they remain for decades. Only when they reach the stratosphere are they decomposed by ultraviolet radiation. Because they accumulate in the atmosphere, CFC concentrations will rise even if emissions are held constant. Figure 9 shows, for CFC 12, the concentrations associated with historical emissions. Note that even as emissions briefly fell in the mid-1970s, concentrations continued to rise.

Projections of Future CFC 11 and CFC 12 Emissions

Quinn et al. (1985) conducted economic analyses and industry interviews to project short-term CFC use. Their baseline results

Figure 6

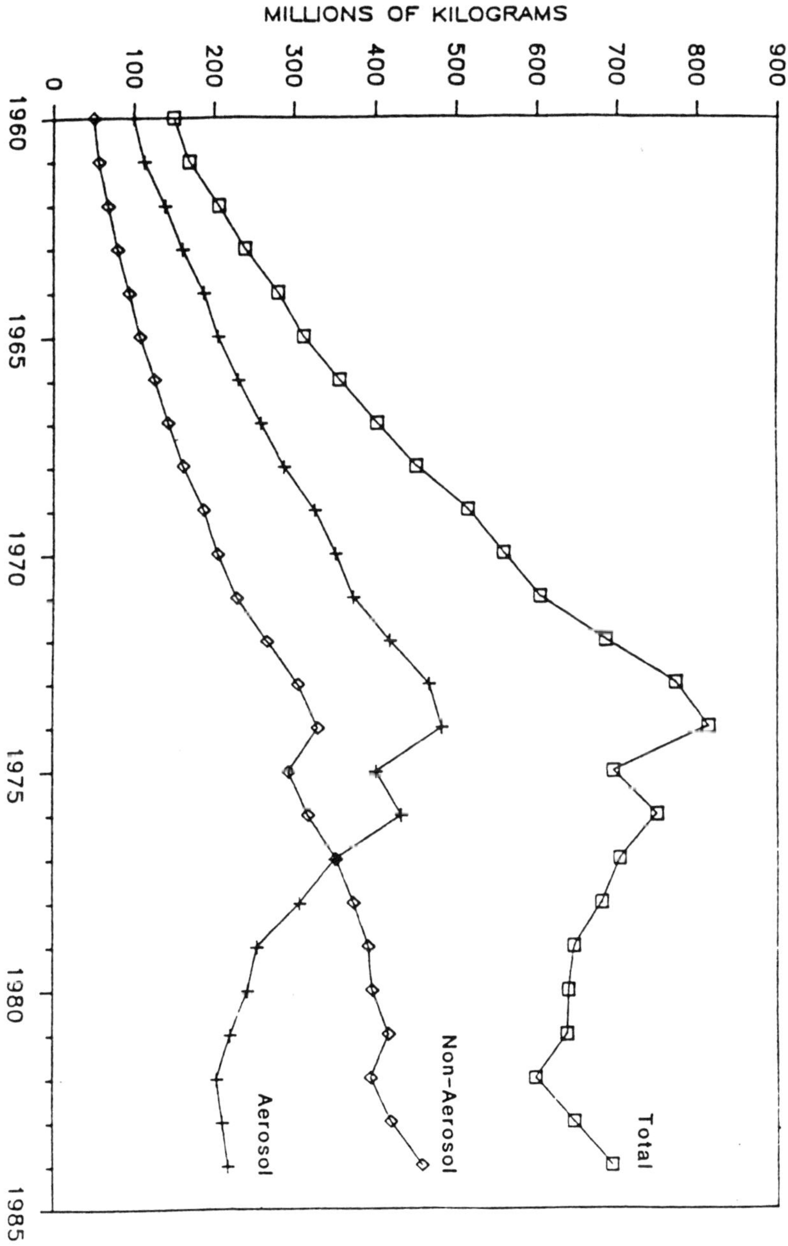

Continued growth of nonaerosol uses has counterbalanced reductions in aerosol uses to produce a relatively constant overall profile of CFC11 and CFC12 use (Chemical Manufacturers Association, 1985).

Figure 7

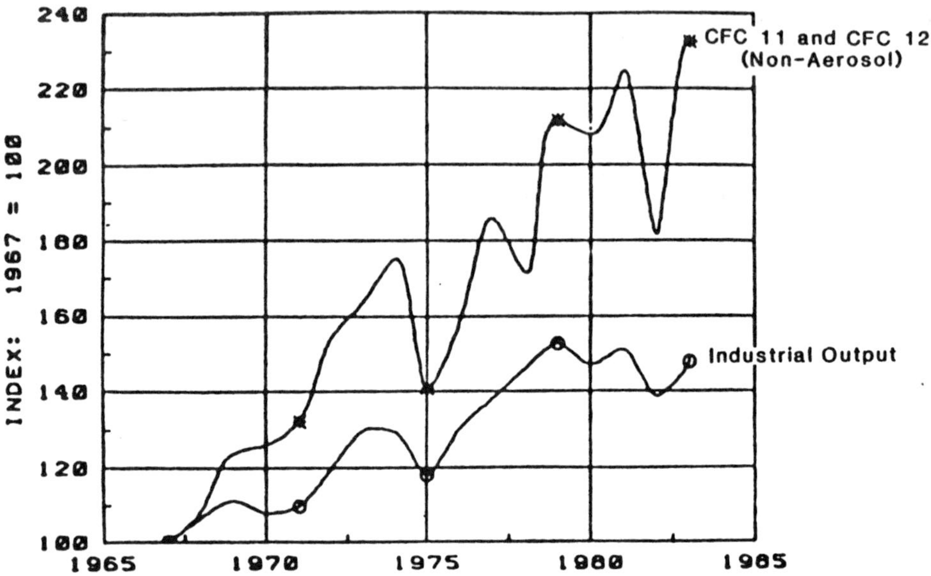

Nonaerosol CFC uses are closely linked to the economy. (Data sources: U.S. International Trade Commission, 1985; Chemical Manufacturers Association, 1985; and Council of Economic Advisers, 1986.)

Figure 8

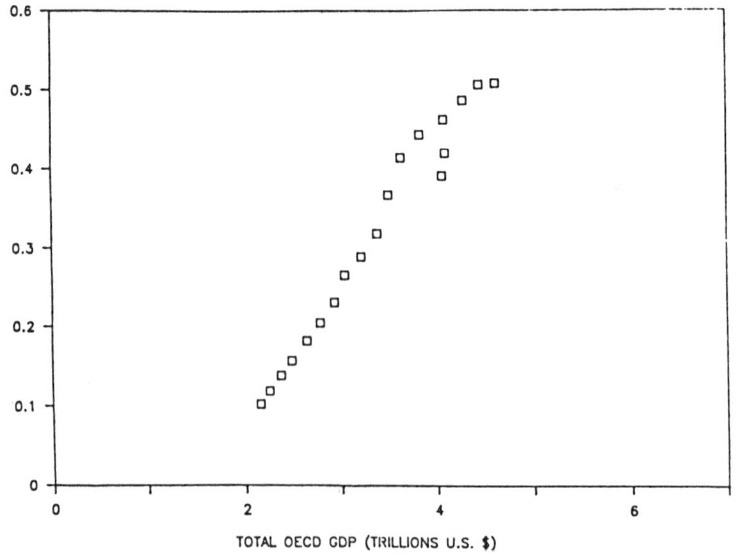

On average, a 1% change in gross domestic product in developed countries is followed by a 2 to 2.5% change in nonaerosol CFC use (Gibbs, 1986).

Figure 9

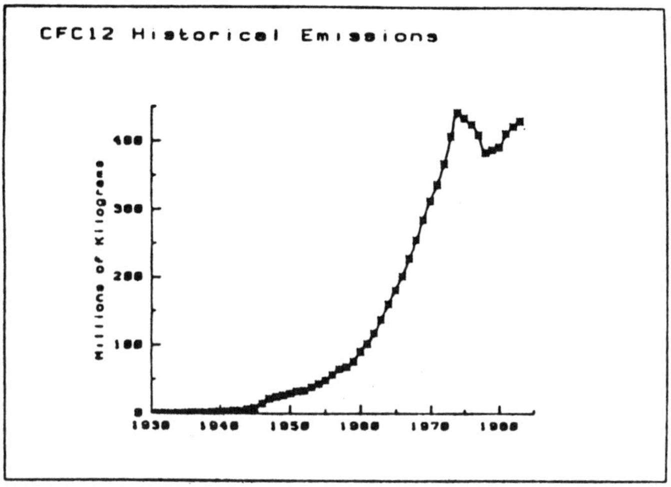

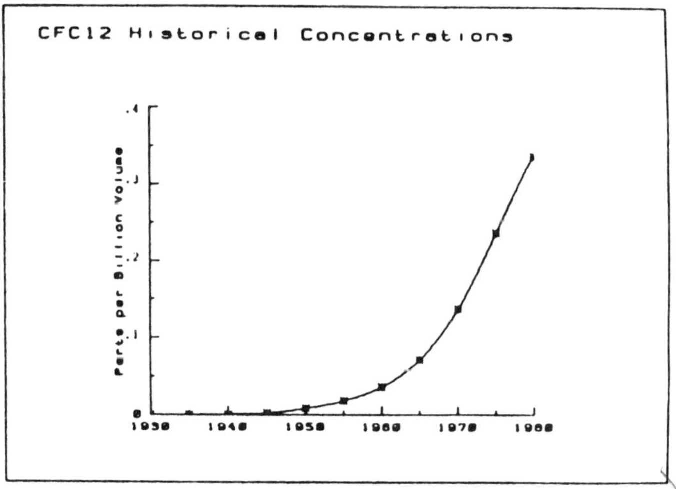

Because they accumulate in the atmosphere, CFC concentrations will rise even if emissions are held constant. Concentrations shown were computed by the authors based on a simplified atmospheric parameterization. Data on emissions are from Chemical Manufacturers Association (1985).

(see Figure 10) show that in the market period of 1985 to 2000, CFC use in the United States would grow at an average annual rate of approximately 4%. While unexpected market forces could, of course, alter this path, these near-term projections should be relatively accurate. This base case represents the consensus of major industry sources in the best position to understand CFC market forces.

Projections further into the future are less certain. For the year 2000 to 2040, Quinn et al. (1985) projected CFC use based on simulations using different assumptions. They investigated the historical growth of CFC demand in the United States and uncovered econometric relations between this demand and population and income. To project future CFC use, Quinn et al. (1985) used assumptions about future population growth (developed by Nathan Keyfritz of Harvard University in conjunction with Edmonds et al., 1984) and future world gross national product (developed by Edmonds et al., 1984). In response to the many uncertainties associated with such a long-term projection, a range of projections was generated (see Figure 10). The most extreme results indicate that during the period 2000 to 2040, world CFC use is likely to grow at an average annual rate of 0.3 to 2.8%. Other researchers, in the United States and abroad, have produced generally consistent estimates of CFC emissions (U.S. Environmental Protection Agency, 1986).

For the extreme long term, 2040 to 2075, we can only consider various contingencies. If current relations between CFC use and world growth continue to hold, Quinn et al. (1985) project CFC increases of 1.0 to 2.1% (see Figure 10). In fact, concentrations of CFCs will rise until the small percentage of CFCs lost from the troposphere to the stratosphere is equal to worldwide emissions, at a date over 100 years off if emissions stopped growing today. Of course, concentrations will rise faster if emissions grow.

Future Projections of CFC 11 and CFC 12 Concentrations

CFC concentrations in any year will depend both on past and present emissions and on future ozone depletion. The ideal model for this effect would consider the time-integrated emissions profile and allow for ultraviolet decomposition in the troposphere equal to that predicted by one- or two-dimensional ozone depletion models. However, such models were unavailable for this analysis, and we computed the concentrations associated with the above scenarios without considering that negative feedback (Figure 11).

Figure 10

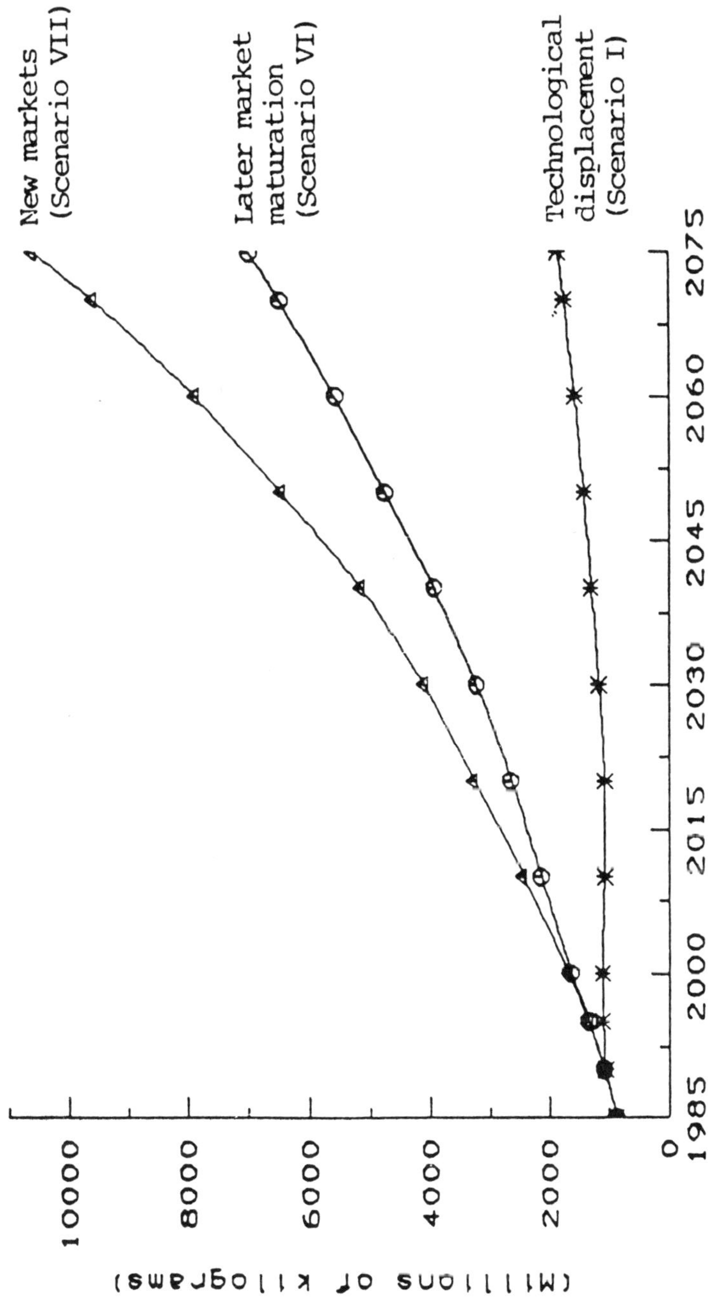

Production estimates from Quinn et al., 1985.

Figure 11

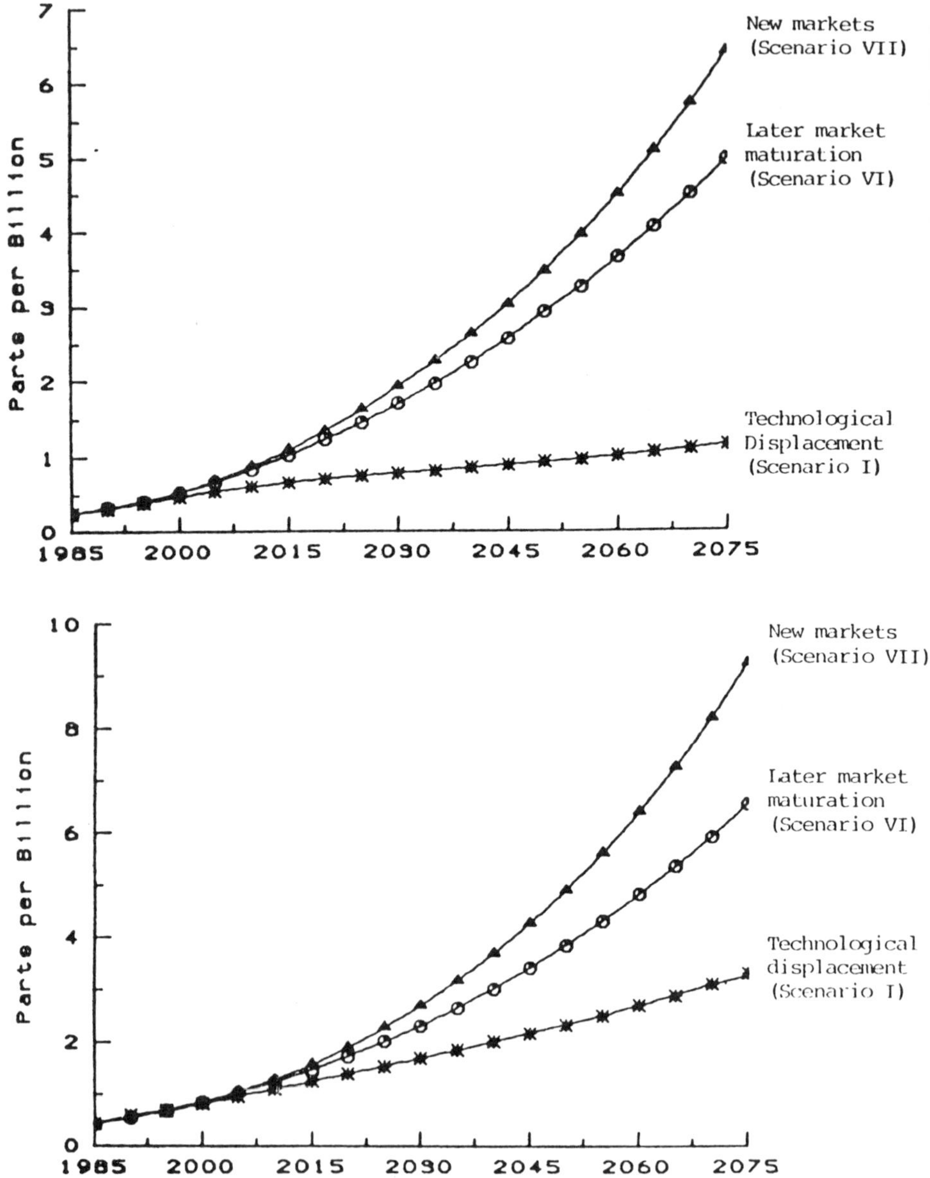

Atmospheric concentrations of CFCs were computed by the authors based on a simplified atmospheric parameterization that does not consider potential feedbacks. The CFC scenarios are taken from Quinn et al. (1985).

METHANE (CH_4)

Measured Increases in Methane Concentrations

The examination of methane bubbles trapped in polar ice cores reveals that methane concentrations had risen by small amounts until approximately 150 to 200 years ago, since when they have roughly doubled. The rate of increase has been escalating, and current increases average about 1% (Rasmussen and Khalil, 1984). Figure 12 shows recent measured concentrations.

Historical Methane Emissions

Rasmussen and Khalil (1984) hypothesize that several hundred years ago, man's influence on the global methane cycle was negligible and natural land-based sources were the dominating factors. This theory is supported by the relatively constant methane concentrations before that time and by the observation that the interhemispheric ratios of dissolved methane in the polar ice cores are consistent with the hemispheric distribution of these sources.

The increase in global methane may be primarily the result of increases in human activities such as rice cultivation, agriculture, raising cattle, and the use of oil and natural gas. The size of each source is, however, uncertain, and it is not yet possible to gauge their relative strengths.

The flux of methane into the atmosphere is not well understood. Table 1 shows a range of estimates of these fluxes. For methane concentrations to grow, either the fluxes would have to increase or the lifetime of methane (estimated as 7 to 11 years) would have to grow.

Methane and the Biogeochemical Cycle

The generation and ultimate fate of methane depend on many physical and chemical processes. Methanogen is produced by anaerobic processes that require biomass and are regulated by temperature and the availability of water. Methane transport to the atmosphere depends on plants and atmospheric conditions. The lifetime of methane, once in the atmosphere, depends on how long it takes for the methane molecules to combine with OH radicals to form other species.

OH availability may be decreasing as a result of increasing levels of carbon monoxide (CO) (Rinsland et al., p. 245 and p. 250, 1985). In fact, Thompson and Cicerone (1985) have demonstrated

Figure 12

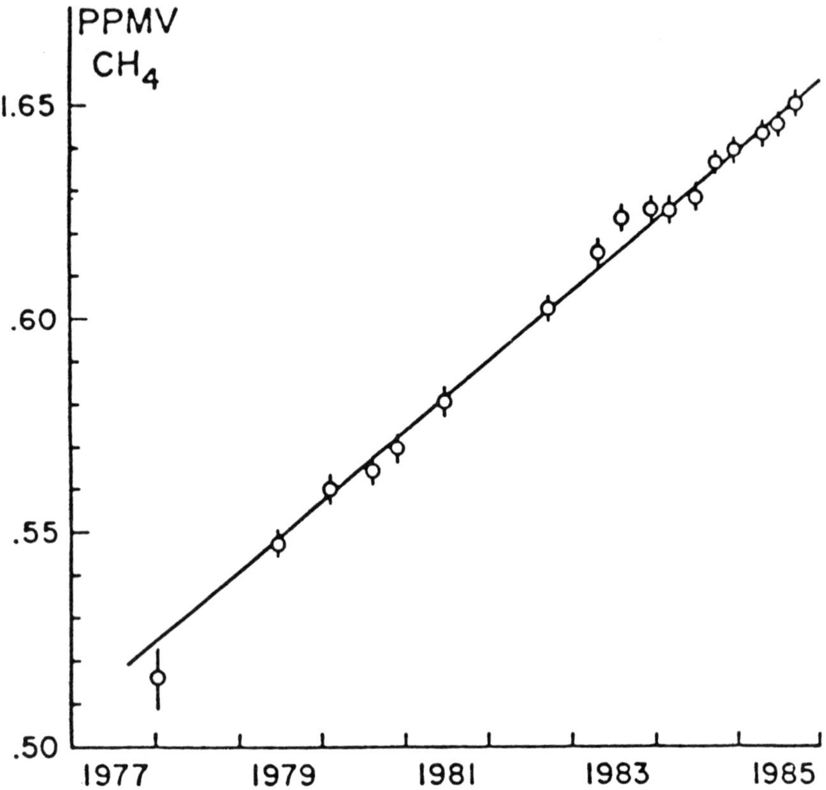

Atmospheric methane concentrations are rising at approximately 1% per year (data from Rowland and Blake, in National Aeronautics and Space Administration, 1986).

Figure 13

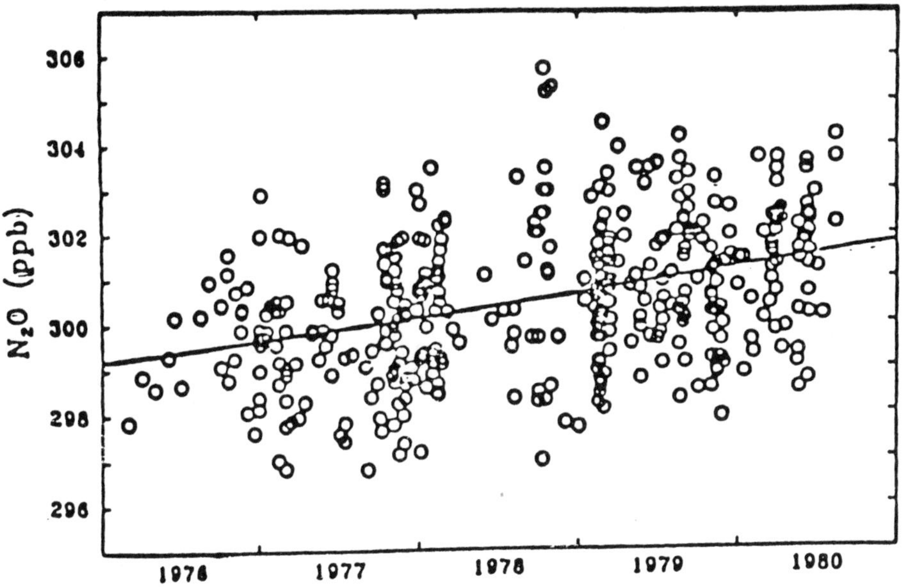

N_2O concentrations are increasing throughout the world (Weiss, 1981).

Table 1
Methane Sources (10^{12} g/year)
(multiply by 0.75 to obtain grams of carbon per year)

Source	I[a]	II[b]
Enteric fermentation (cattle, sheep, etc.)	100-200	100-150
Rice paddies[c]	app. 280	100± 50
Wetlands[d]	190-300	150± 50
Biomass burning[e]	---	10- 60
Freshwater lakes[f]	1-25	---
Oceans[g]	1-17	---
Tundra[f]	0.3-3	---
Anthropogenic / fossil fuel[g]	16-50	---
Other[h]	---	10-150
Total	586-825	390-765

[a]Column I is taken from Enhalt (1974) and Enhalt and Schmidt (1978).
[b]Column II is taken from Khalil and Rasmussen (1983).
[c]All rice studies.
[d]Area estimate from Harris and fluxes from Khalil and Rasmussen.
[e]$\Delta CH_4/\Delta CO_2$ = 0.3-2%, Seiler et al. paper.
[f]Assumed in footnote d.
[g]Should be updated.
[h]Seiler et al.

that rising CO emissions could be the chief cause of the rise in CH_4, not increasing CH_4 emissions. The complexities of all the processes make it impossible to know.

Future Concentrations of Methane

To accurately predict the concentrations of methane would require understanding many things, including how the sources of methane fluxes will change over time (how much more riceland, cattle, how much less wetland); how emission rates per relevant unit (e.g., acre of rice) will change due to alterations in temperature, water, cultivation practices, and even carbon dioxide; how CO emissions will change; how tropospherical chemical balances will change. Current science is not yet capable of making estimates for all these questions. Thus we must fall back to extrapolating concentrations--1% per year. In reality, methane rises may tail off or rise, depending on how all these processes unfold.

NITROUS OXIDE (N_2O)

Measured Increases in N_2O Concentrations

Analysis of air samples collected from 1976 to 1980 by Weiss (1981) shows that concentrations of N_2O have recently risen at about 0.2% per year (Figure 13). Recent data from five measuring stations around the world confirm this trend (National Aeronautics and Space Administration, 1986).

Historical N_2O Emissions

The major sources of N_2O are fossil fuel combustion and fertilizer denitrification. Researchers estimate that fossil fuel combustion currently accounts for between 20 and 30% of emissions, with fertilizers and natural sources contributing the remainder.

N_2O and the Biogeochemical Cycle

The only known sink for N_2O is stratospheric photolysis and reaction with singlet atomic oxygen. Both processes occur only in the upper atmosphere, and the atmospheric residence time of N_2O is consequently long: approximately 100 to 175 years. As with the CFCs, this means that even a decrease in N_2O emissions may not achieve a decrease in N_2O concentrations.

Figure 14

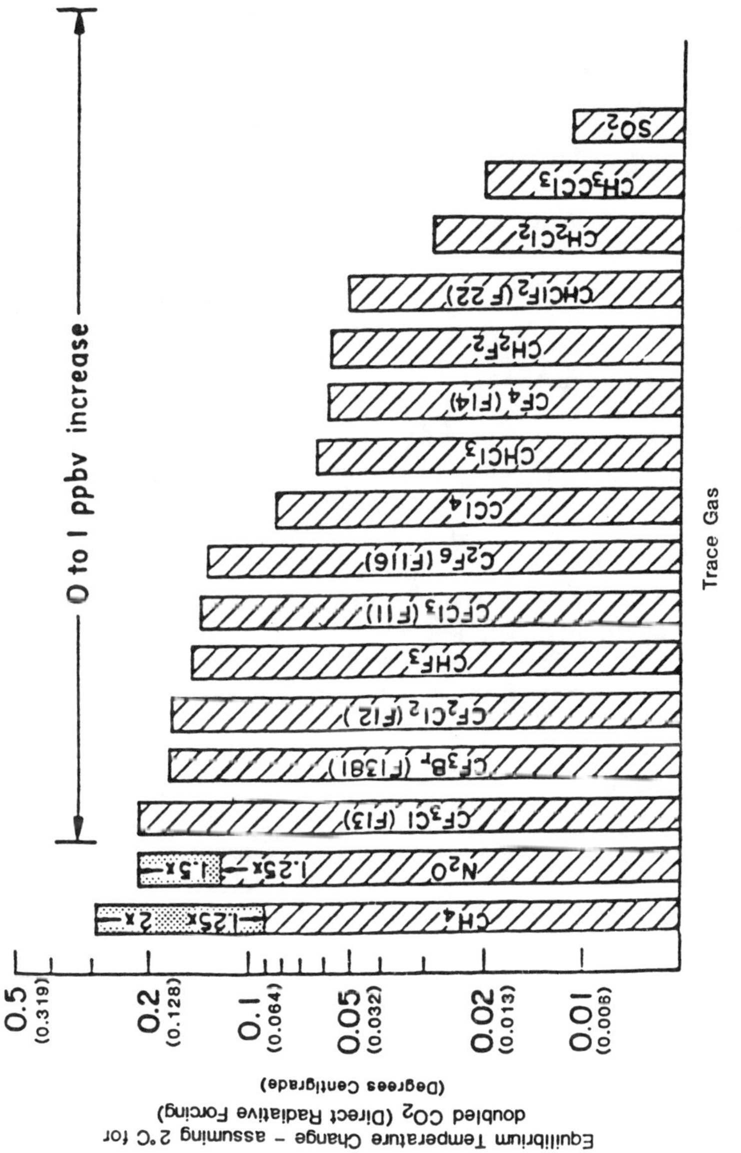

The relative greenhouse strengths of trace gases are shown in terms of the equilibrium temperature change for a temperature sensitivity of $2^{0}C$ ($3.6^{0}F$), and the direct radiative forcing independent of climatic feedbacks. Direct radiative forcing is shown in parentheses. Graph is adapted from Ramanathan et al. (1985).

Projections of Future N_2O Emissions and Concentrations

Future emissions of N_2O will primarily depend on fossil fuel combustion, agricultural activity, and other natural sources. Efforts are currently under way to estimate N_2O emission scenarios from long-term energy growth. Estimates of future changes in the flux of agricultural and natural emission have not yet been developed, however.

Estimates of N_2O growth for use in scenario testing in the literature include those of Ramanathan et al. (1985), who project increases of 0.3%, and Wuebbles et al. (1984), who suggest annual increases of 0.0 to 0.3%.

OTHER TRACE GASES

While carbon dioxide, the chlorofluorocarbons, methane, and nitrous oxide are expected to contribute the most to global warming, researchers have identified a score of minor trace gases which are radiatively active. Such gases include nitrogen compounds, sulfur compounds, fully fluorinated species, chlorocarbons, brominated and iodated species, hydrocarbons, aldehydes, and tropospheric ozone. Based on a radiative forcing model with an overall sensitivity of 2°C (3.6°F) for doubled CO_2, Ramanathan et al. find that by 2030, the combined effect of these minor trace gases should be approximately 0.05°C (0.09°F).

CONCLUSION

Emissions of greenhouse gases and their atmospheric concentrations have been rising for decades. The release of these gases to the atmosphere is intimately related to sustained economic growth. Figure 14 shows the relative strength of each gas per part per billion (ppb) for 2°C (3.6°F) sensitivity and the radiative forcing of each gas, which is independent of any feedbacks. Clearly any analysis of a global warming must encompass much more than the use of CO_2.

REFERENCES

Broecker, W. S., and T. Takahashi. 1983. Is there a tie between atmospheric CO_2 content and ocean circulation? Presented at Symposium on Ice and Climate Modeling, Evanston, Ill.

Brown, S., and A. E. Lugo. 1981. The role of the terrestial biota in the global carbon cycle. *Div. Petrol. Chem.* 26, 1019–1025.

Callendar, G. S. 1958. On the amount of carbon dioxide in the atmosphere. *Tellus* 10, 243-248.

Chemical Manufacturers Association. 1985. *Production, Sales, and Calculated Release of CFC-11 and CFC-12 through 1984.* Chemical Manufacturers Association, Washington, D.C.

Clarke, W. C. (ed.). 1982. *Carbon Dioxide Review: 1982.* Oxford University Press, New York.

Council of Economic Advisers. 1986. *Economic Report of the President*, U.S. Government Printing Office, Washington, D. C.

Council on Environmental Quality. 1981. *Global Energy Futures and the Carbon Dioxide Problem.* U.S. Government Printing Office, Washington, D.C.

Delmas, R. J., J. M. Ascensio, and M. Legrand. 1980. Polar ice evidence that atmospheric CO_2 20,000 B.P. was 50% of present. *Nature* 284, 155-157.

Edmonds, J. A., J. Reilley, J. R. Trabalka, and D. E. Reichle. 1984. *An Analysis of Possible Future Atmospheric Retention of Fossil Fuel* CO_2. U.S. Department of Energy, Washington, D.C.

Emanuel, W., G. Kilbugh, and J. Olson. 1981. Modeling the circulation of carbon in the world's terrestial ecosystems. *In* B. Bolin. et al. (ed.), *Modeling the Global Carbon Cycle--SCOPE 16.* John Wiley & Sons, Inc., New York.

Freyer, H. D. 1978. Preliminary evaluation of past CO_2 increase as derived from 13C measurements in tree rings. *In* J. Williams (ed.), *Carbon Dioxide, Climate and Society.* Pergamon Press, New York.

Gibbs, M. J. 1986. *Scenarios of CFC Use: 1985-2075.* ICF Inc., Washington, D.C.

Houghton, R. A., et al. 1983. Changes in the carbon content of terrestial biota and soils between 1860 and 1980: a net release of CO_2 to the atmosphere. *Ecol. Monogr.* 53, 235-262.

Lacis, A., J. Hansen, P. Lee, T. Mitchell, and S. Lebedeff. 1981. Greenhouse effect of trace gases, 1970-1980. *Geophysical Res. Lett.* (8)10, 1035-1038.

National Aeronautics and Space Administration. 1986. Tropospheric trace gases: sources, distributions and trends. *In Present State of Knowledge of the Upper Atmosphere: Processes That Control Ozone and Other Climatically Important Trace Gases.* National Aeronautics and Space Administration, Washington, D.C.

Neftel, A., H. Oeschger, J. Schwander, B. Stauffer, and R. Zumbruun. 1982. Ice core sample measurements give atmospheric CO_2 content during the past 40,000 years. *Nature* 295, 220-223.

Nordhaus, W. D., and G. W. Yohe. 1983. Future paths of energy and carbon dioxide emissions. *In Changing Climate.* National

Academy Press, Washington, D.C.

Oeschger, H., B. Stauffer, A. Neftel, J. Schwander, and R. Zumbruun. 1982. Atmospheric CO_2 content in the past deduced from ice-core analyses. *Ann. Glaciol.* 3, 227-232.

Peng, T. H., W. S. Broecker, H. D. Freyer, and S. Trumbore. 1983. A deconvolution of the tree ring based $d_{13}C$ record. *J. Geophys. Res.* 88, 3609-3620.

Quinn, T. H., K. A. Wolf, W. E. Mooz, J. K. Hammitt, T. W. Chesnutt, and S. Sarma. 1985. *Projected Use, Emissions, and Banks of Potential Ozone Depleting Substances.* Draft report. The RAND Corp., Santa Monica, Calif.

Ramanathan, V., H. B. Singh, R. J. Cicerone, and J. T. Kiehl. 1985. Trace gas trends and their potential role in climate change. *J. Geophys. Res.* 90(D3), 5547-5566.

Rasmussen, R. A., and M. A. K. Khalil. 1984. Atmospheric methane in the recent and ancient atmospheres: concentrations, trends, and interhemispheric gradient. *J. Geophys. Res.* 90(D3), 5547-5566.

Rinsland, C. P., J. S. Levine, and T. Miles. 1985. Free tropospheric carbon monoxide concentrations in 1950 and 1951 deduced from infrared total column amount measurements. *Nature* 318, 250-253.

Rinsland, C. P., J. S. Levine, and T. Miles. 1985. Concentration of methane in the troposphere deduced from 1951 infrared solar spectra. *Nature* 318, 245-249.

Rotty, R. M., and G. Marland. 1984. *The Changing Pattern of Fossil Fuel CO_2 Emissions.* DOE/OR/21400-2. U.S. Department of Energy, Washington, D.C.

Seidel, S. and D. Keyes. 1983. *Can We Delay a Greenhouse Warming?* GPO/055-000-00235-5. U.S. Government Printing Office, Washington, D.C.

Thompson, A. M., and R. J. Cicerone. 1985. Possible perturbations to atmospheric CO_2, CH_4, and OH. Accepted by *J. Geophys. Res.*

Tyndall, J. 1861. On the absorption and radiation of heat by gases and vapours, and on the physical connexion of radiation, absorption, and conduction. *Philosoph. Magazine* (22)146, 169-194.

U.S. Environmental Protection Agency. 1986. *Summary Paper for Topic #2: Projections of Demand for CFCs.* Prepared for United Nations Environment Programme Workshop in the Ozone Layer, Rome, Italy.

U.S. International Trade Commission. 1985. *Report on U.S. Production of Selected Synthetic Organic Chemicals (Including Plastics and Resin Materials).* U.S. International Trade Commission,

Washington, D.C.

Weiss, R. F. 1981. The temporal and spatial distribution of tropospheric nitrous oxide. *J. Geophys. Res.* 86(C8), 7185-7195.

Wuebbles, D. J., M. C. MacCracken, and F. M. Luther. 1984. *A Proposed Reference Set of Scenarios for Radiatively Active Atmospheric Constitutents.* DOE/NBB-0066. U.S. Department of Energy, Washington, D.C.

FORESTS--
THE MOMENTUM FOR A GREENHOUSE EFFECT: PHYSICAL, ECONOMIC, AND POLITICAL LIMITS TO CURTAILING THE MAGNITUDE OF GLOBAL WARMING

by John S. Hoffman and John Bruce Wells

ABSTRACT

Governments may wish to adopt policies to limit greenhouse warming. Scenarios are presented that show the effects of successfully implementing such policies and that evaluate the difficulties introduced by physical, economic, and political obstacles.

Physical obstacles include the world's oceans, which delay its approach to thermal equilibrium, and the long lifetime of many greenhouse gases, which makes it difficult to control their concentrations. Economic obstacles consist of the many strong links between greenhouse gas emissions and economic and population growth. This relationship may be difficult to dislodge. The political obstacles to coordinated international action are also large. Policies will be costly to implement and slow to work.

Two sets of policies are tested: halving the growth rate of greenhouse gas emissions and decreasing them by 1% per year. Over the next 50 years, even the successful adoption of a policy as strict as a 1% annual decrease in emissions cannot prevent a temperature increase that equals or exceeds that of the last century. Choice of a policy makes a significant difference in reducing long-term global warming.

Before foresters can decide how to respond to changes anticipated as a result of a changing atmosphere, they must examine whether the warming will be limited by policies implemented by government. This paper examines that possibility, evaluating the obstacles to significantly limiting the warming in the next 50 years and the 50 years thereafter. Three sets of difficulties in limiting global warming will be analyzed:

- physical obtacles,
- economic obstacles, and
- political obstacles.

Scenarios will be presented which show the effects of successfully initiating various policies to limit or reduce emissions of "greenhouse gases" for a range of uncertainties about the temperature sensitivity of the earth.

PRIOR STUDIES

Several studies have evaluated the world's ability to delay or limit global warming. Perry et al. (1982) show the relationship between action initiation time and the severity of the policy needed to achieve various goals. Their analysis was centered on CO_2 alone and did not consider other trace gases. Nor did they use an atmospheric model capable of estimating change in equilibrium temperatures and the delays caused by ocean heat absorption and, thus, the actual temperature change likely to be experienced in any year. Nevertheless, their analysis showed one thing very clearly: the later action is initiated, the more radical the policy must be to achieve a given CO_2 limit.

Seidel and Keyes (1983), in a widely reported EPA study, analyzed the effects of various energy policies including a 100% and 300% U.S. tax on fossil fuels, a 100% and 300% world tax on fuels producing CO_2, and a coal and synfuels ban. All were initiated in 2000. Their study included a coupled atmospheric and oceanic model that allows estimation of actual transient temperature, and they considered the warming contribution of methane, nitrous oxide, and chlorofluorocarbons (CFCs), but not the possibility of controlling them. They showed that fossil fuel policies alone have remarkably small effects. Assuming the midrange assumption of the National Academy of Sciences of 3°C (5.4°F) for a doubling of CO_2, they found that a 300% tax could delay a 2°C warming by only 5 years.

Lovins et al. (1981) argue that potential improvements in energy efficiency and renewable energy resources will allow the world to stabilize and eventually eliminate fossil fuel use without dampening economic and population growth. This policy would limit global warming more than estimated by Seidel and Keyes. However, since the publication of the Lovins report, energy use has not fallen as suggested. In fact, the recent fall in energy prices is likely to spur energy growth.

BASELINE SCENARIOS ASSUMED IN THIS ANALYSIS

The scenarios in this analysis are derived from a variety of sources which are discussed in our previous paper. Table 1 summarizes the projected increases in radiatively active trace gases and their basis in the literature.

Table 1
Projected Increases in Radiatively Active Trace Gases: Independent Estimates and Assumptions of This Study

A. Carbon Dioxide

	Emissions (GT/year)						Year of Doubled CO_2 Concentrations
	1975	2000	2025	2050	2075	2100	(300 to 600 ppm)
NAS (1983)							
Low (25th percentile)	5	4	7	8	10	12	2100
High (75th percentile)	5	7	13	17	22	27	2050
Seidel & Keyes (1983)							
Low demand	5	7	8	10	19	33	2073
High fossil	5	7	11	16	35	59	2059
Edmonds et al. (1984)							
Low (Case C)	5	6	7	7	7	NA	after 2100
High (Case A)	5	8	19	47	91	NA	(2033 to 2039)

This study: Low case emissions from NAS 25th percentile; high from NAS 75th percentile. Our carbon cycle model produces CO_2 doubling times of 2056 and 2091.

B. Methane

	Concentrations (Annual % Growth)
Ramanathan et al. (1984)	
Low (1980 to 2030)	0.25
High (1980 to 2030)	1.50
Khalil and Rasmussen (1985)	
Low (1980 to 2000)	0.97
High (1980 to 2000)	1.10
This study	
Low	0.70
High	1.40

C. Nitrous Oxide

	Emissions (Annual % Growth)	Concentrations (Annual % Growth)
Weiss (1981)		
Current trends 2020	NA	0.20
Low (1980 to 2020)	3.5	0.40
High (1980 to 2020)	6.0	0.70
This study		
Low	1.0	0.20
High	1.5	0.25

D. CFCs	CFC 11 and CFC 12 Use (Annual % Growth)			Concentrations (ppbv)						
				2000		2030		2060		
	1980-2000	2000-2030	2030-2060	CFC 11	CFC 12	CFC 11	CFC 12	CFC 11	CFC 12	
Quinn et al.										
Slow growth (#2)	2.3	1.0	1.0	0.5	0.8	1.1	1.8	1.8	3.0	Emissions modeled to concentrations in this study
New markets (#7)	4.1	3.0	2.2	0.6	0.9	2.1	2.9	4.8	6.8	
Gibbs (1985)										
Low	1.2	1.6	1.8	NA	NA	NA	NA	NA	NA	
High	10.6	10.3	0.2	NA	NA	NA	NA	NA	NA	
Ramanathan et al. (1984)										
Low	NA	NA	NA	NA	NA	0.5	0.9	NA	NA	
High	NA	NA	NA	NA	NA	2.0	3.5	NA	NA	

This study: Low case emissions from RAND Scenario 2; high from Rand Scenario 7.

E. Minor Trace Gases	Ramanathan et al. (1984)				Gibbs (1985) Emissions (1980 to 2060) (Annual % Growth)		Quinn et al. (1985) Emissions (1980 to 2060) (Annual % Growth)		This Study Concentration in 2030 (ppbv)		
	1980 Conc.	2030 Conc. (ppbv)									
Gas	(ppbv)	Low	Best	High	Low	High	Low	High	Low	High	
SO_2	0.1	0.1	0.1	0.2	NA	NA	NA	NA	0.1	0.2	
CF_4	0.07	0.2	0.24	0.31	NA	NA	NA	NA	0.24	0.31	
C_2F_6	0.004	0.01	0.02	0.04	NA	NA	NA	NA	0.02	0.04	Concentrations extrapolated
CH_2Cl_2	0.03	0.1	0.2	0.3	NA	NA	NA	NA	0.2	0.3	
$CClF_3$	0.007	0.04	0.06	0.1	NA	NA	NA	NA	0.06	0.1	
$CHCl_3$	0.01	0.02	0.03	0.1	NA	NA	NA	NA	0.03	0.1	
C_2H_2	0.06	0.1	0.1	0.16	NA	NA	NA	NA	0.1	0.16	
CFC 22	0.06	0.4	0.9	1.9	0.9	3.0	NA	NA	0.19	0.75	Emissions modeled to concentrations in this study
CCl_4	0.13	0.2	0.3	0.4	NA	NA	0.05	3.05	0.14	0.34	
CH_3CCl_3	0.14	0.7	1.5	3.7	NA	NA	0.80	2.85	0.23	0.62	
Halon 1301	0.001	0.003	0.005	0.01	NA	NA	3.71	6.47	0.01	0.02	

Global emissions sources are uncertain for SO_2. CF_4. C_2F_6. CH_2Cl_2, $CClF_3$. $CHCl_3$. and C_2H_2. We adopt the estimates of Ramanathan et al. of their 2030 concentrations. In the low case we hold the concentrations constant after 2030; in the high case we extrapolate.

Our estimates of the future concentrations of CFCs, CCl_4. CH_3CCl_3. and Halon 1301 do not consider ozone depletion. which may allow increased penetration of ultraviolet radiation which dissociates these compounds and would thereby lower their concentrations.

In this study, the assumed equilibrium sensitivity to a doubling of CO_2 concentrations is 2°C and 4°C. When we consider other radiatively active gases. the effective CO_2 doubling date advances to 2053 (low case) and 2024 (high case).

To estimate the warming associated with rises in greenhouse gases we use the atmospheric model developed by Lacis et al. (1981), modified to include the minor trace gases identified by Ramanathan et al. (1985). Several gases and feedback effects that will increase temperature rise are omitted, including the water vapor that will occur in the stratosphere as methane increases, and increases in surface temperature due to changes in the vertical distribution of stratospheric ozone.

The warming ultimately associated with a given increase in greenhouse gases is called the "equilibrium warming." The equilibrium temperature, however, is not immediately realized in the year the gas increases because additional radiation prevented from escaping the earth must raise the temperature of the upper layers of the ocean in tandem with atmosphere. The Lacis model includes a "box model" representation of oceans that allows consideration of the delay caused by oceanic heat absorption. Thus the model produces output that includes both the global temperature actually realized in each year (i.e., the transient temperature) and the equilibrium temperature that will eventually be achieved without any increase in the concentrations of greenhouse gases. From these outputs it is possible to derive the unrealized warming, i.e., warming that is already in the system, but not yet experienced.

Figure 1 shows the realized and equilibrium warming expected for low and high assumptions about the earth's temperature sensitivity and increases in radiatively active trace gases.

THE EFFECTS OF VARIOUS POLICIES AND THEIR INITIATION TIMES

Two sets of "policies" to limit or reduce global warming are examined:

- Halving the worldwide growth rate of emissions: the growth rates for the emissions of CO_2, N_2O, CFCs, CH_3CCl_3, and Halon 1301, and the growth rate of atmospheric concentration of methane in 2000. For the other minor trace gases discussed by Ramanathan et al. (1985), the atmospheric concentrations are leveled.[1]

---

[1]Given the long atmospheric residence times and scattered sources of many of these other minor trace gases, leveling their concentrations is likely to be a more stringent policy than halving their emissions and may be equivalent to an emissions decrease.

Figure 1
Baseline Scenarios of Global Warming

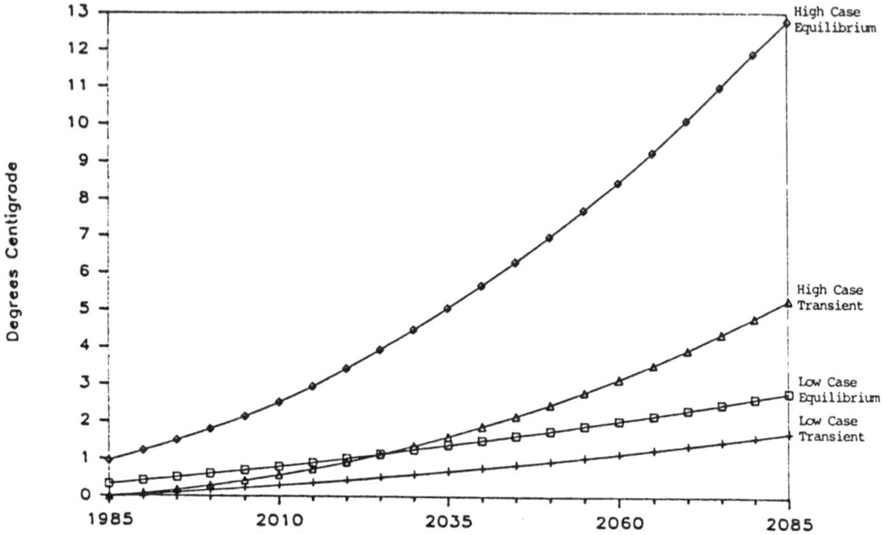

The realized and equilibrium warming is shown for low and high assumptions about the earth's temperature sensitivity and future increases in greenhouse gases.

Figure 2
Global Warming in the Next 50 Years

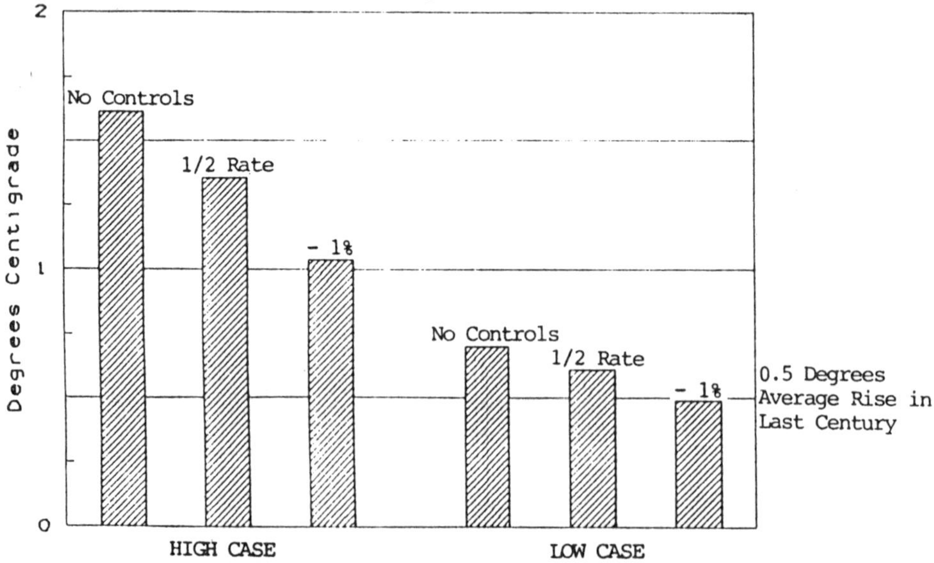

The figure shows that realized warming in the next 50 years exceeds that of the last 100 years for all three policies considered regardless of the true sensitivity of the climate system.

- Decreasing emissions by 1% per year: the emissions of CO_2, N_2O, CFCs, CH_3CCl_3, and Halon 1301 and the atmospheric concentration of methane are cut by 1% per year. The atmospheric concentrations of the other minor trace gases are held constant.

If no policies are adopted, as Figures 1 and 2 show, in the next 50 years the temperature increase will far exceed the average temperature rise of 0.5°C (0.9°F) of the past century. Depending on growth in trace gases and the earth's thermal sensitivity, the increase will range from 0.7 to 1.6°C (1.2 to 2.9°F). (The paper of Lacis et al., 1981, projects an even larger change.)

Over the next 50 years, even the adoption of control policies as severe as a 1% annual decrease in emissions cannot prevent a temperature increase which equals or exceeds that 0.5°C rise of an entire century. Figure 2 shows that halving the growth rate cannot prevent a temperature increase of 0.6 to 1.4°C (1.1 to 2.5°F) and decreasing emissions by 1% per year cannot prevent an increase of 0.5 to 1.0°C (0.9 to 1.8°F).

By the year 2085, the policy adopted makes a significant difference in global warming. Figures 1 and 3 show that if no policies are adopted, mankind may experience unprecedented temperature increases: 1.7 to 5.3°C (3.1 to 9.5°F). Action by 2000 to cut emissions growth in half would reduce this warming to 1.3 to 3.5°C (2.3 to 6.3°F). Action to decrease emissions by 1% per year, starting in 2000, would reduce the warming to 0.8 to 1.8°C (1.4 to 3.2°F).

The differences in the relative effectiveness of these policies in reducing long-term global warming are even more striking if we consider unrealized warming in 2085. In 2085, even if emissions or changes in the concentration of gases were completely terminated, the earth's temperature would continue to rise as the oceans and atmosphere approached radiative equilibrium. If no policies are adopted until then to control gases, this additional, unpreventable warming (in addition to realized warming) would be 1.2 to 8.6°C (2.2 to 15.5°F) (Figure 4). Cutting emissions growth in half would reduce this unrealized warming to 0.8 to 4.8°C (1.4 to 8.6°F); cutting emissions by 1% per year would reduce it to 0.3 to 1.6°C (0.5 to 2.9°F).

One thing is particularly noticeable from this analysis: greenhouse warming in the next several decades is not terribly responsive to uncertainties about either the earth's temperature sensitivity or policies to reduce warming. Regardless of these uncertainties, over the next 50 years the earth's warming will match or exceed that of the past 100 years. Over the much longer term,

Figure 3
Global Warming in the Next 100 Years

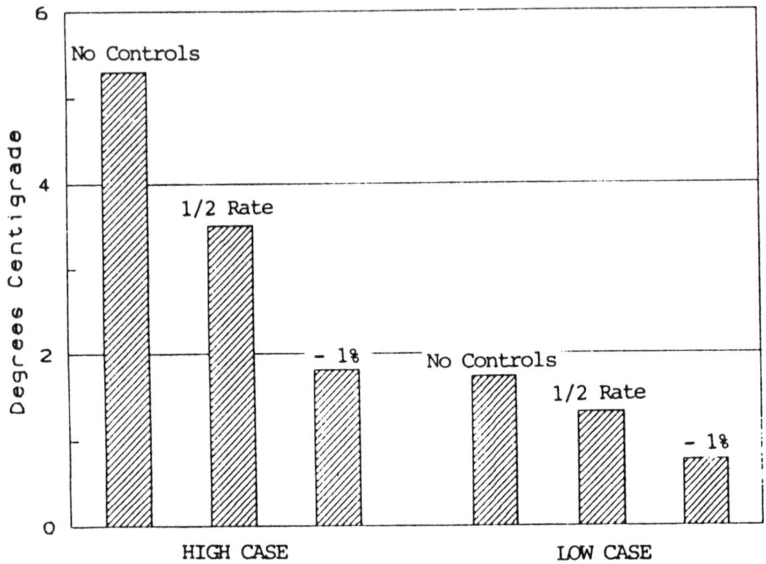

The figure shows that by the year 2085, the policy selected of those considered has a large effect, but realized temperature change is still significant.

Figure 4
Unrealized Warming in the Next 100 Years

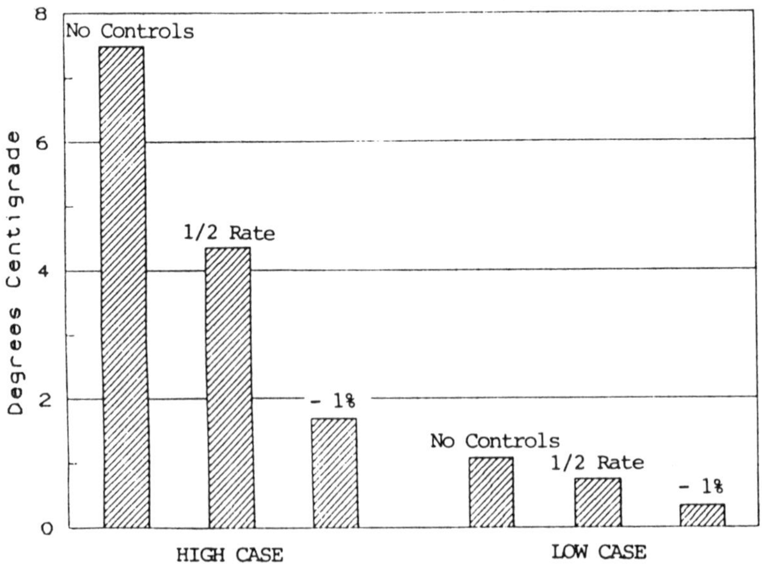

Even if trace gas concentrations did not change after the year 2085, additional and unpreventable warming would still occur. Policies could, however, significantly reduce the magnitude of this unrealized warming.

from 2035 to 2085, control policies have markedly different effects. The reason for both conclusions is the same: the momentum of the overall system. There are two sources of this momentum. First, the delays in experiencing warming from gases already emitted, which may have benefited us, also encumber us with unrealized change which policies taken soon or in the future cannot prevent. The second source of momentum is biogeochemical--that is, since many greenhouse gases have long atmospheric lifetimes, the atmosphere is currently far out of balance, so that emissions entering it far exceed current losses from it, leading to a rapid rise in concentration. For such gases, curtailing emissions growth will only slow their rise. Like any system that has slowly built up speed and momentum, it will take time to slow the system's change.

ECONOMIC AND POLITICAL LIMITS

The policies discussed, halving or decreasing emissions by 2000, would require nations to take steps such as bans or fees or traditional regulations to accomplish their goals. As Edmonds and Reilly (1985) demonstrate, this will not be easy; they find that "even under an admittedly unrealistic scenario of global cooperation in taxing CO_2 emissions (at a 100% rate), the doubling window (for CO_2 concentrations) shifts less than a decade." In 1980 RAND Corporation estimated the costs of controls to reduce CFC emissions (Figure 5). RAND found that even raising the price of CFCs by 300% could not level emissions; they projected a reduction of growth from 5.4% per year to 2.6%. Methods to curtail emissions of methane and nitrous oxide are less certain. For example, controlling methane may be possible by switching rice crops or by controlling CO emissions, but at this time it is impossible to predict accurately the effects of such policies. Almost no research has been done on methods of curtailing any emissions but CO_2 and CFCs.

Successful policies must be implemented worldwide. Seidel and Keyes (1983), for example, showed that for CO_2, policies in the United States alone would be ineffective because the price of fuel on world markets would drop, thereby increasing demand. Regardless of which general policy approach is under consideration, any hope of success requires cooperation among at least the major energy-consuming and energy-producing nations. Without worldwide cooperation, international fuel trading and unconstrained use of fuel in noncooperating countries would impede the effectiveness of any policy. For CFCs such price considerations are less important

Figure 5
CFC Control Costs

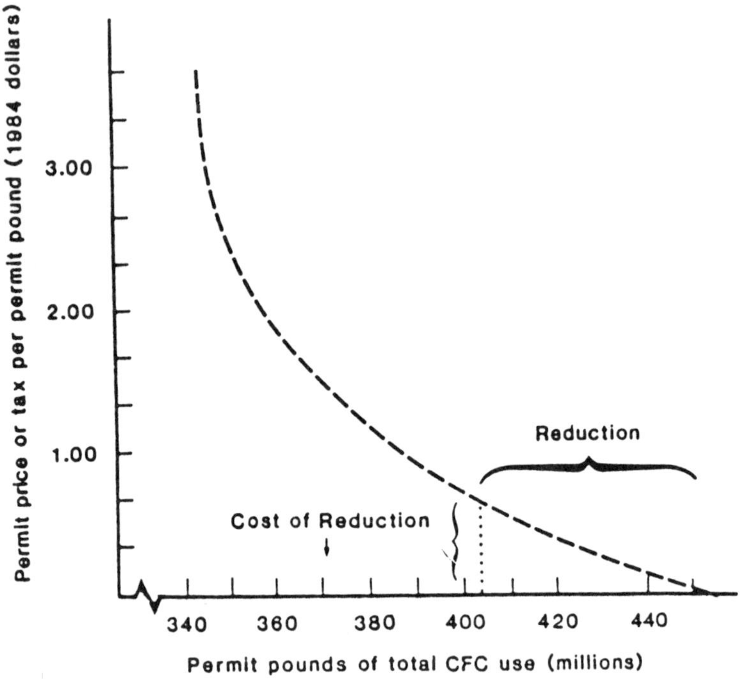

RAND found that increases in the prices of CFCs would not reduce demand very much with currently available technologies (RAND, 1980).

(because reducing U.S. demand does not lower world CFC prices appreciably). International cooperation, however, would strengthen the effectiveness of CFC control policies. The U.S. aerosol ban undertaken in 1978 reduced CFC emissions and therefore global warming. Nevertheless the failure of much of the world to join this ban diminished the total reduction that could have been made. Had most of the world followed the United States, the reduction would have been greater.

In the real world, policies will be costly to implement and slow to work. Even the enormous oil price hikes of the 1970s only slightly decoupled energy use from economic growth. Given these facts, it seems very unlikely that significant global warming during the near future can be prevented.

If one accepts the consensus that has emerged in the scientific community, foresters are very likely to face significant climatic warming in coming decades regardless of the policies governments adopt to limit it. The magnitude of warming that will occur will depend on how world governments react to the challenge, but warming will almost certainly continue to alter the environment for decades thereafter.

REFERENCES

Edmonds, J., and J. M. Reilly. 1985. *Global Energy: Assessing the Future*. Oxford University Press, New York, N.Y.

Lacis, A., J. Hansen, P. Lee, T. Mitchell, and S. Lebedeff. 1981. Greenhouse effect of trace gases. *Geophys. Res. Lett.* 8: 1035-1038.

Lovins, A. B., L. H. Lovins, F. Krause, and W. Bach. 1981. *Least Cost Energy: Solving the CO_2 Problem*. Brick House Publishing Co., Inc., Andover, Mass.

National Academy of Sciences. 1983. *Changing Climate*. National Academy Press, Washington, D.C.

Perry, A. M., K. J. Araj, W. Fulkerson, O. J. Rose, M. M. Miller, and R. M. Rotty. 1982. Energy supply and demand implications of CO_2. *Energy* 7(12), 991-1004.

Ramanathan, V., H.B. Singh, R. J. Cicerone, and J. T. Kiehl. 1985. Trace gas trends and their potential role in climate change. *Geophys. Res.* 90(03), 5547-5566.

RAND. 1980. *Economic Implications of Regulating Chlorofluorocarbon Emissions from Nonaerosol Applications*. The RAND Corporation, Santa Monica, Calif.

Seidel, S., and O. Keyes. 1983. *Can We Delay a Greenhouse Warming?* GPO #055-000-00235-5. U.S. Government Printing Office, Washington, D.C.

Appendix A
Low Case

	No Controls (°C)			Control Policies: Halved Growth Rate of Emissions (°C)			Control Policies: 1% Annual Decrease in Emissions (°C)		
Year	Equilibrium Temperature	Transient Temperature	Unrealized Warming	Equilibrium Temperature	Transient Temperature	Unrealized Warming	Equilibrium Temperature	Transient Temperature	Unrealized Warming
1985	0.337	0.000	0.337	0.337	0.000	0.337	0.337	0.000	0.337
1990	0.433	0.057	0.376	0.433	0.057	0.376	0.433	0.057	0.376
1995	0.527	0.117	0.410	0.527	0.117	0.410	0.527	0.117	0.410
2000	0.618	0.178	0.440	0.618	0.178	0.440	0.618	0.178	0.440
2005	0.713	0.242	0.471	0.696	0.239	0.457	0.670	0.234	0.436
2010	0.814	0.308	0.506	0.776	0.299	0.477	0.721	0.284	0.437
2015	0.921	0.379	0.542	0.857	0.359	0.498	0.768	0.331	0.437
2020	1.034	0.454	0.580	0.938	0.421	0.517	0.810	0.375	0.435
2025	1.152	0.532	0.620	1.022	0.483	0.539	0.848	0.417	0.431
2030	1.272	0.615	0.657	1.105	0.547	0.558	0.882	0.456	0.426
2035	1.392	0.700	0.692	1.188	0.612	0.576	0.913	0.493	0.420
2040	1.515	0.788	0.727	1.271	0.677	0.594	0.941	0.528	0.413
2045	1.641	0.879	0.762	1.355	0.744	0.611	0.966	0.560	0.406
2050	1.770	0.973	0.797	1.440	0.812	0.628	0.988	0.591	0.397
2055	1.903	1.070	0.833	1.527	0.880	0.647	1.008	0.621	0.387
2060	2.042	1.171	0.871	1.614	0.950	0.664	1.026	0.648	0.378
2065	2.187	1.275	0.912	1.704	1.022	0.682	1.041	0.674	0.367
2070	2.337	1.384	0.953	1.795	1.095	0.700	1.055	0.699	0.356
2075	2.493	1.498	0.995	1.888	1.169	0.719	1.067	0.722	0.345
2080	2.652	1.615	1.037	1.981	1.245	0.736	1.078	0.743	0.335
2085	2.814	1.736	1.078	2.075	1.323	0.752	1.088	0.764	0.324
2090	2.978	1.861	1.117	2.170	1.401	0.769	1.097	0.784	0.313
2095	3.145	1.988	1.157	2.266	1.480	0.786	1.106	0.802	0.304
2100	3.314	2.119	1.195	2.363	1.561	0.802	1.114	0.820	0.294

Assumes 2°C sensitivity for doubled CO_2 and low growth in radiatively active trace gases.

Appendix B
High Case

Year	No Controls (°C)			Control Policies: Halved Growth Rate of Emissions (°C)			Control Policies: 1% Annual Decrease in Emissions (°C)		
	Equilibrium Temperature	Transient Temperature	Unrealized Warming	Equilibrium Temperature	Transient Temperature	Unrealized Warming	Equilibrium Temperature	Transient Temperature	Unrealized Warming
1985	0.967	0.000	0.967	0.967	0.000	0.967	0.967	0.000	0.967
1990	1.228	0.089	1.139	1.228	0.089	1.139	1.228	0.089	1.139
1995	1.506	0.192	1.314	1.505	0.192	1.314	1.506	0.192	1.314
2000	1.807	0.308	1.499	1.807	0.308	1.499	1.807	0.308	1.499
2005	2.145	0.438	1.707	2.084	0.432	1.652	1.990	0.422	1.568
2010	2.527	0.584	1.943	2.380	0.563	1.817	2.169	0.533	1.636
2015	2.952	0.749	2.203	2.688	0.704	1.984	2.330	0.640	1.690
2020	3.423	0.934	2.489	3.009	0.853	2.156	2.477	0.745	1.732
2025	3.936	1.140	2.796	3.344	1.012	2.332	2.610	0.846	1.764
2030	4.481	1.367	3.114	3.685	1.179	2.506	2.731	0.944	1.787
2035	5.058	1.615	3.443	4.032	1.354	2.678	2.840	1.038	1.802
2040	5.665	1.882	3.783	4.384	1.537	2.847	2.939	1.130	1.809
2045	6.306	2.170	4.136	4.743	1.726	3.017	3.029	1.218	1.811
2050	6.984	2.479	4.505	5.110	1.924	3.186	3.111	1.303	1.808
2055	7.701	2.810	4.891	5.485	2.128	3.357	3.185	1.385	1.800
2060	8.459	3.164	5.295	5.857	2.340	3.527	3.253	1.464	1.789
2065	9.260	3.542	5.718	6.256	2.559	3.697	3.314	1.540	1.774
2070	10.105	3.945	6.160	6.653	2.786	3.867	3.368	1.614	1.754
2075	10.997	4.375	6.622	7.057	3.020	4.037	3.418	1.686	1.732
2080	11.899	4.829	7.070	7.462	3.260	4.202	3.463	1.754	1.709
2085	12.788	5.301	7.487	7.864	3.507	4.357	3.506	1.821	1.685
2090	13.666	5.787	7.879	8.262	3.757	4.505	3.547	1.885	1.662
2095	14.532	6.285	8.247	8.658	4.012	4.646	3.585	1.948	1.637
2100	15.389	6.793	8.596	9.053	4.271	4.782	3.622	2.010	1.612

Assumes 4°C sensitivity for doubled CO_2 and high growth in radiatively active trace gases.

EVIDENCE FOR FUTURE WARMING: HOW LARGE AND WHEN?

by James Hansen, Andrew Lacis, David Rind, Gary Russell, Inez Fung, and Sergej Lebedeff

ABSTRACT

A wide range of evidence confirms a general understanding of the "greenhouse" mechanism. Climate models and empirical evidence from past climate changes suggest a climate sensitivity for doubled atmospheric carbon dioxide (CO_2) that agrees with the estimate of 3 ± 1.5 °C (5.4 ± 2.7 °F) by committees of the National Academy of Sciences. Recent rapid increases of atmospheric trace gases also have accelerated the greenhouse forcing of the climate system. An experiment, using a three-dimensional global climate model in which CO_2 and other trace gases increase at the rate observed from 1958 to 1984 and at similar rates into the future, shows global temperatures inching above the level of natural variability by about 1990, followed by a trend of rapidly rising temperatures.

Forest managers and climate modelers must work together to design and implement models that are useful for forestry applications. Managers can help modelers decide which climate variables most influence forests, and modelers can instruct managers on the capabilities and limitations of model-generated climatic change scenarios. This paper is intended to begin that interdisciplinary discussion.

The "greenhouse effect" is the warming of the earth's surface caused by atmospheric gases that absorb infrared radiation emitted by the earth. The atmosphere radiates some of this energy back to the ground, thus warming the surface, as a blanket retains thermal warmth. A quantitative estimate of the greenhouse effect follows from the observation that the temperature in the earth's atmosphere falls off at a rate of about 6°C per km (7°F per mile) as altitude increases (Figure 1).

Figure 1
Greenhouse Effect

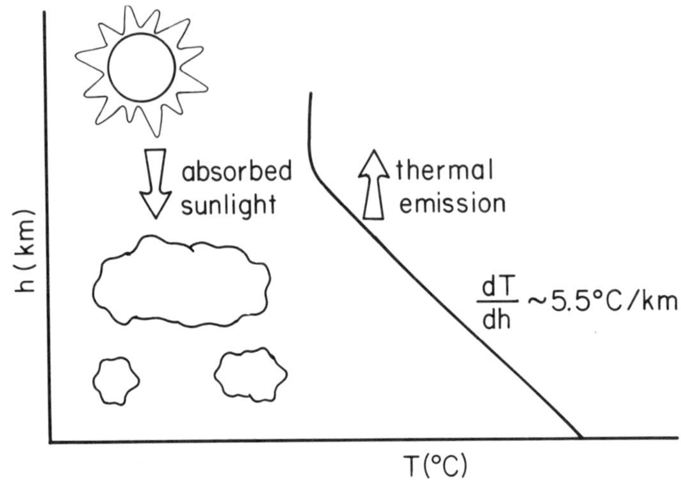

Absorbed sunlight must be balanced by emitted thermal radiation, in the long run. Today the mean level of emission to space is from an altitude of about 6 km; thus the surface temperature (15°C or 59°F) is about 33°C (59°F) warmer than the mean radiating temperature of the earth (-13°C or 9°F), because temperature falls off with increasing altitude by about 5.5°C per km (6.2°F per mile). If atmospheric CO_2 were doubled, the atmosphere would become more opaque in the infrared, raising the mean level of emission to space by about 200 m. If everything (including the lapse rate, dT/dh) remained fixed, i.e., if there were no climate feedbacks, the temperature would have to increase about 1.2°C (2.1°F) with doubled CO_2 to restore energy balance with space.

Table 1
Greenhouse Effect on Terrestrial Planets

							T_s(K)		
Planet	S_0 (W m^{-2})	A	τ	Γ(°C km^{-1})	H (km)	T_e (K)	Eq. (1)	Eq. (2)	Observed
Mars	589	0.15	0.1	5.0	1	217	221	222	220
Earth	1367	0.30	1.0	5.5	6	255	293	288	288
Venus	2613	0.75	100.0	7.0	70	232	685	720	700

T_e is expected planetary temperature in the absence of a greenhouse effect. The expected temperature including greenhouse warming is computed from two different simple approximations, equations (1) and (2). The observed surface temperatures are given in the final column. Equation (1): Radiative equilibrium $T_s = T_e\ (1 + 3/4(\tau))^{1/4}$. Equation (2): Convective equilibrium $T_s = T_e + \Gamma H$. Symbols: S_0 = solar irradiance; A = planetary albedo; τ = atmospheric infrared opacity; Γ = atmospheric mean lapse rate; H = mean altitude of emission to space; T_e = effective temperature = $S_0(1 - A/4^o)^{1/4}$; T_s = surface temperature.

At any given wavelength (λ), energy is emitted to space from a range of altitudes centered at the level from which atmosphere's optical depth (τ_λ) is unity, as measured from above. If an infrared-absorbing (or greenhouse) gas is added to the atmosphere, infrared atmospheric opacity increases and raises the mean level of emission to space to a greater altitude, where it is colder. For the standard test case of doubled atmospheric carbon dioxide (CO_2) (from, say, 300 to 600 parts per million [ppm]), the mean level of emission to space is computed to approximately 200 m. Thus, if everything including the lapse rate remained fixed, the atmosphere and surface temperature would have to increase by about 1.2°C (2°F) for the thermal emission from the planet to remain in balance with the amount of absorbed solar radiation. In reality, everything else will not remain fixed, since a warmer climate is likely to cause, for example, an increase in atmospheric water vapor and a decrease in sea ice cover. The observed temperatures on the three terrestrial planets, Mars, Earth, and Venus, elucidate the greenhouse mechanism. As indicated in Table 1, the amount of solar energy absorbed by these planets is similar, so their temperatures would all be in the range of 215 to 255 Kelvin (-72 to 0°F) in the absence of a greenhouse effect. Because of the large range of infrared atmospheric opacity on the three planets, the expected greenhouse effects are a few degrees on Mars, a few tens of degrees on Earth, and several hundred degrees on Venus, as computed with simple greenhouse models. The agreement of observed warmings on these planets with the theoretical expectations--for a case in which the planets have had time to reach equilibrium and in which feedbacks affecting atmospheric composition or planetary albedo have been explicitly incorporated--supports the greenhouse theory.

CLIMATE SENSITIVITY ESTIMATES

A more detailed estimate of climate sensitivity on Earth can be obtained from global climate models (GCMs) and from empirical evidence provided by past climate (paleoclimate and historical climate) change.

Climate Models

At least three GCMs have been used to compute the equilibrium climate response to doubled atmospheric CO_2. The Geophysical Fluid Dynamics Laboratory (GFDL) model (Manabe and Stouffer, 1980) computes a global warming of about 2°C (3.6°F), whereas the Goddard Institute for Space Studies (GISS) model (Hansen et al., 1983 and 1984b, hereafter referred to as papers 1 and 2) and the National Center for Atmospheric Research (NCAR) model (Washington and Meehl, 1984) yield about 4°C (7.2°F).

The geographical distribution of the annual mean surface-air warming in the GISS model is shown in Figure 2. There is a high-latitude enhancement of the warming caused by the positive ice and snow albedo feedback and by the relative stability of the atmosphere at high latitudes, which tends to concentrate the warming in the lowest layers of the atmosphere. High-latitude enhancement of the warming is also found in the GFDL and NCAR models.

The different physical processes contributing to the 4°C (7.2°F) mean warming in the GISS model have been analyzed by inserting the changes obtained in the three-dimensional model one by one into a one-dimensional radiative/convective climate model. As shown in the lower part of Figure 2, the net effect of the changes in atmospheric water vapor is a large positive feedback with feedback factor (f_w) of ~1.6. Decreases in ground albedo as a result of reduced ice and snow cover yield a feedback factor (f_i) of ~1.1; changes in cloud height and cloud cover yield f_c ~1.3. Positive feedbacks are mutually reinforcing; thus, as the formalism in paper 2 (Hansen et al., 1984b) indicates, the net for these three processes in the GISS model is a feedback factor (f) of 3 to 4.

These analyses help explain the differences in sensitivities obtained with various climate models and identify physical processes that contribute most to uncertainty about true climate sensitivity. For example, the principal difference between the GISS and GFDL models is in f_c, the assumption of fixed clouds in the GFDL model implying $f_c = 1$. Present information on clouds is inadequate to evaluate cloud prediction schemes, making clouds a major reason for the uncertainty in climate sensitivity. The International Satellite Cloud Climatology Project (Schiffer and Rossow, 1983) is expected to improve understanding of the influence of clouds on climate sensitivity.

Because of the uncertainties in climate models, especially cloud feedback, current models only roughly estimate that global mean climate sensitivity is 3°C (5.4°F) for doubled atmospheric CO_2 (*see* Charney, 1979, and Smagorinsky, 1982; the "error bar" they give [±1.5°C or 2.7°F] is very subjective). This pure-model result is then quantitatively compared with empirical data from climate changes in the past, as discussed below.

Figure 2

GLOBAL WARMING IN 3-D MODEL FOR DOUBLED CO_2

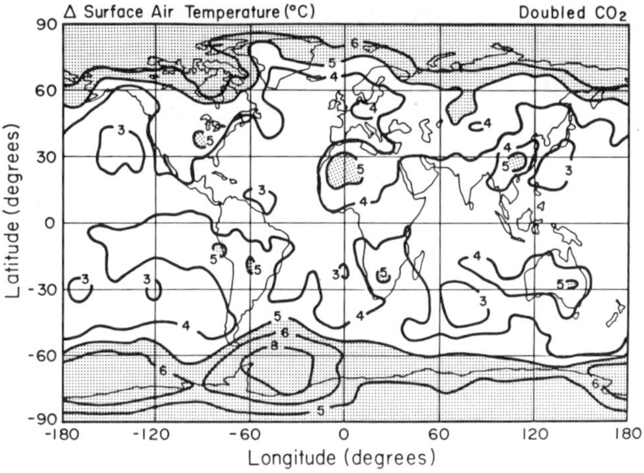

CONTRIBUTIONS TO GLOBAL WARMING IN MODEL

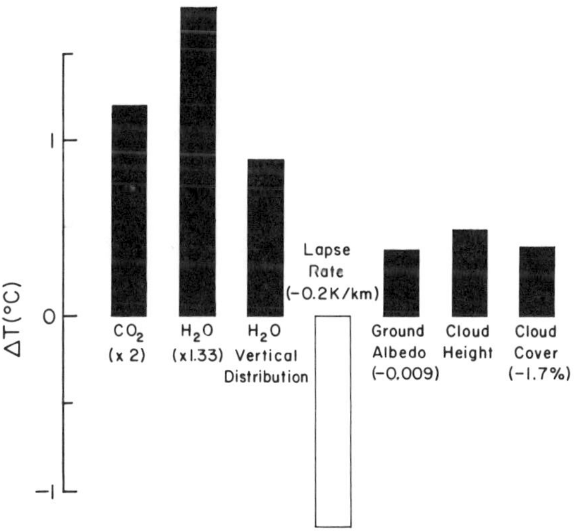

Annual mean warming obtained in a three-dimensional global climate model (GISS model, described by Hansen et al. in papers 1 and 2 (1983 and 1984b)) at equilibrium ($t \rightarrow \infty$) for doubled atmospheric CO_2. Lower part of figure shows contributions of different physical processes to the global warming, as estimated by inserting the changes in the three-dimensional model into a one-dimensional radiative/convective model (see Hansen et al., paper 2, 1984b).

Last Ice Age

Climatic boundary conditions such as ice coverage and ocean-surface temperature have been reconstructed in detail for the Wisconsin Ice Age (18,000 years ago) by the CLIMAP project (CLIMAP, 1981). In a general circulation model these boundary conditions can accurately determine the global surface-air temperatures of the Wisconsin Ice Age. As shown in Figure 3, the mean temperature in the Wisconsin Ice Age was about 4°C (7.2°F) colder than today, with the greatest cooling at and near the ice sheets over North America and Europe.

The basic mechanisms for glacial to interglacial climatic changes are thought to be very subtle, such as the small changes in radiative forcing caused by variation of the earth orbital elements (Milankovitch, 1941). However, it is not necessary to unravel the sequence of cause and effect of the ice ages to estimate climate sensitivity from the known boundary conditions of the last ice age. Hansen et al. (1984a) show that the 4°C (7.2°F) cooling of the last ice age is consistent with the climate forcing computed from the CLIMAP boundary conditions, if climate sensitivity is in the range 2 to 5°C (4 to 9°F) for doubled CO_2.

Information from a past ice age confirms the approximate magnitude of the equilibrium climate sensitivity estimated directly from climate models. The paleoclimate approach should improve evaluations of climate sensitivity as knowledge of paleoclimate boundary conditions improves and climate models become more realistic.

Last Century

Atmospheric CO_2 abundance has increased significantly in the past century, a period for which measurements of surface-air temperature are available. The observed temperature change is compared with the documented CO_2 change to obtain an empirical measure of climate sensitivity. Any conclusion is limited by the implicit assumption that 100-year trends of other climate forcings (such as solar irradiance or aerosols, for example) are small compared with the greenhouse effect of CO_2 change.

This empirical test does not compare equilibrium climate states. Because of the significant heat capacity of the ocean, the response time of the surface-air temperature to a change in climate forcing is of the same order of magnitude as the characteristic time for change of CO_2 (much of the CO_2 change occurring, for example, in the past 25 years). To properly incorporate the ocean

Figure 3

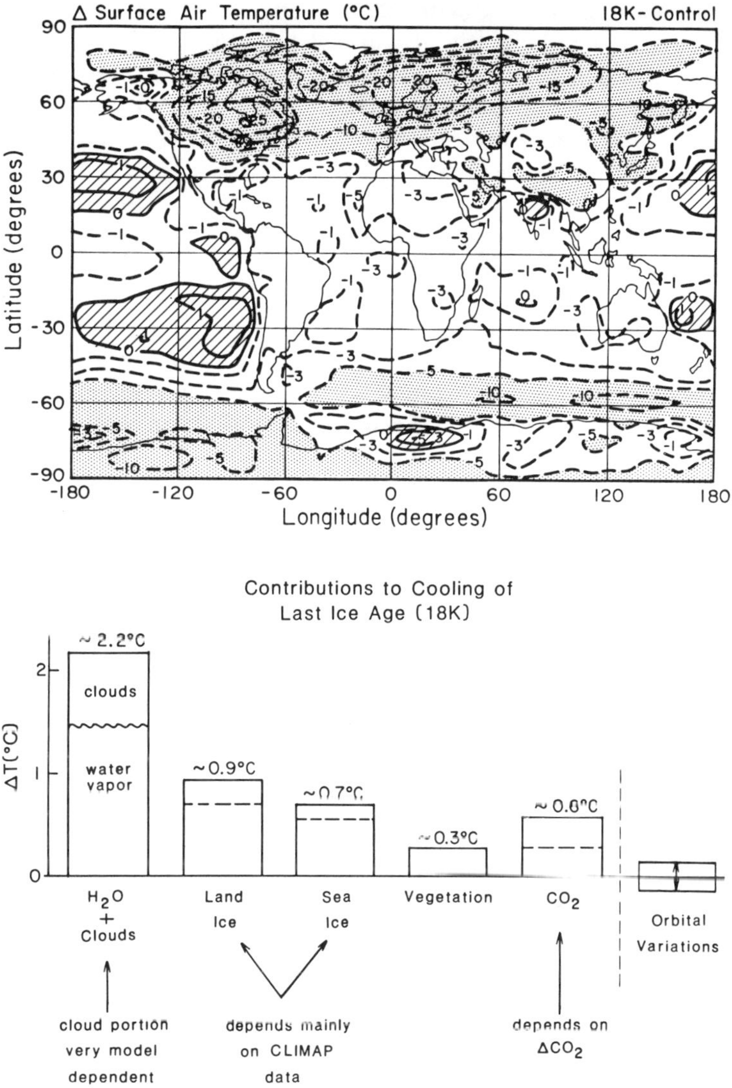

Annual mean surface-air temperature difference between the climates of today and the Wisconsin ice age (18,000 years ago; 18K), based on three-dimensional climate model simulations with the CLIMAP boundary conditions (Hansen et al., 1984b). Areas in which the cooling exceeded 5°C (9°F) are stippled, while areas with warmer temperatures at 18K are indicated by slashes. The contributions of different physical processes to the global mean cooling at 18K are shown in the lower part of the figure, as estimated by inserting individual changes of boundary conditions into the climate model.

response, a realistic ocean-climate model that includes the dependence of ocean circulation on climate change at the ocean surface must be constructed. In the absence of such ocean-modeling capability, available information on ocean mixing rates from transient ocean tracers (tritium and chlorofluorocarbons, for example) is the best substitute, which requires the assumption that a *small* heat perturbation mixes as a passive tracer.

The expected amount of warming between 1850 and 1980 is shown in Figure 4 as a function of the climate sensitivity and the 1850 CO_2 abundance. A CO_2 warming of 0.5°C (0.9°F) implies a climate sensitivity of ~4°C (7°F) for doubled CO_2, if the 1850 CO_2 abundance were 270 ppm. Uncertainty in the actual 1850 to 1980 degree of warming increases the range of acceptable climate sensitivities. It is apparent from Figure 4 that the lower limit on climate sensitivity [$\Delta T_{eq}(2^*CO_2)$] is much sharper than the upper limit. It is not possible to push the lower limit beneath about 2°C (3.6°F) for the likely ranges of 1850 CO_2 abundance and temperature. On the other hand, if the 1850 CO_2 abundance were as high as 280 ppm (Oeschger, 1984), then any uncertainty in the observed warming implies that no practical upper limit on climate sensitivity can be established by this method.

The commonly assumed empirical temperature increase of 0.5°C (0.9°F) for the period 1850 to 1980 suggests a climate sensitivity similar to that indicated by climate models and by paleoclimate data. The precision of the climate sensitivity inferred from recent climatic trends is limited by uncertainties about past atmospheric composition, the true global mean temperature change and its cause, and the manner and rate at which the ocean takes up heat.

TRACE GAS TRENDS

Greenhouse forcing of the climate system has accelerated recently because several trace gases have increased (Figure 5). When Keeling began monitoring CO_2 in 1958, its abundance was 315 ppm. Thus, if the 1850 CO_2 abundance was 250 ± 10 ppm as estimated by the World Meteorological Organization (WMO) (1983), the mean decadal growth of CO_2 for the period 1850 to 1958 was about 4 ppm per decade. By contrast, the current rate of growth of CO_2 is three to four times greater.

Chlorofluorocarbons

Moreover, the growth of other gases now adds a greenhouse forcing comparable to that of CO_2. The major source of these other

Figure 4

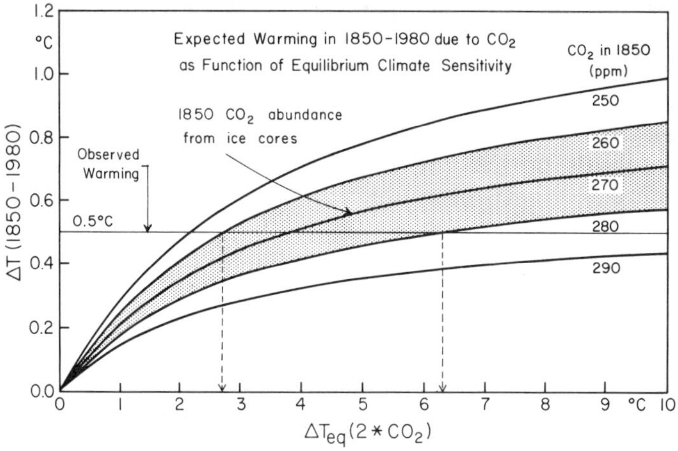

Computed global warming between 1850 and 1980 as a function of the equilibrium climate sensitivity for doubled CO_2. The shaded area is the range of CO_2 abundance in 1850 estimated by the World Meteorological Organization (1983).

Figure 5

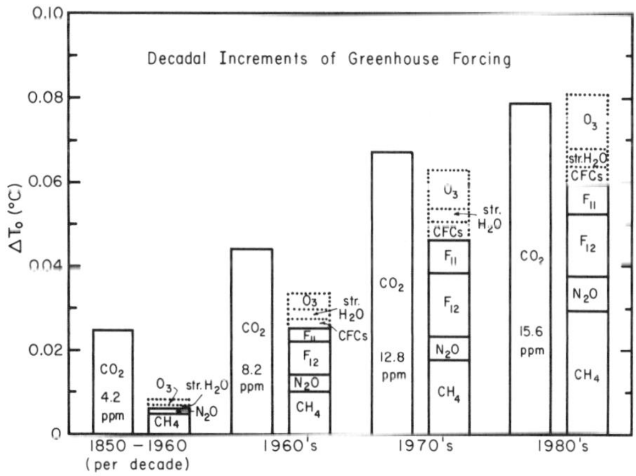

Decadal additions to global mean greenhouse forcing of the climate system. ΔT_0 is the computed temperature change at equilibrium ($t \rightarrow \infty$) for the estimated decadal increases in trace gas abundances, with no climate feedbacks included. Multiply ΔT_0 by the feedback factor f to get the equilibrium surface temperature change including feedback effects. Most of the estimated trace gas increases are based on measurements, as discussed in the text. The O_3 and stratospheric H_2O trends (dotted bars) are based principally on model calculations (Wuebbles et al., 1983).

gases is chlorofluorocarbons. The release rates of CCl_2F_2 (trade name Freon 12 or F_{12}) and CCl_3F (F_{11}) are known quite reliably; their greenhouse forcings are shown individually in Figure 5. Less abundant chlorine and fluorine compounds, such as $CHClF_2$ (F_{22}), CCl_4, and CF_4, have significant growth rates and infrared absorption strengths. A probable contribution from the other chlorofluorocarbons is on the order of 10 to 20% that of CCl_2F_2 and CCl_3F.

Methane

Methane has been measured for the last several years, but it is uncertain whether its increase is confined to recent years or has been going on for several centuries, as suggested by some measurements of air bubbles trapped in polar ice cores (Craig and Chou, 1982). The reality of the early CH_4 increase (before 1965) has been questioned (Ehhalt et al., 1983); however, even if the rate of increase of CH_4 obtained by Craig and Chou (1982) for the period before 1958 is correct, its contribution to the greenhouse forcing of the climate system is small (Figure 5).

Increase of methane may also cause an indirect greenhouse effect: an increase of stratospheric water vapor, which is produced by methane oxidation. The greenhouse effect of the added stratospheric H_2O, which Wuebbles (private communication, 1984) estimates to be associated with increase of CH_4, is almost half of the direct CH_4 greenhouse warming.

Nitrous Oxide

Nitrous oxide increased slightly in the 1970s, at a rate of about 0.2% per year (Weiss, 1981). Recent measurements (Carbon Dioxide Assessment Committee, 1983) suggest a current growth rate of about 0.3% per year.

Ozone

Ozone is an effective greenhouse material because of its strong absorption band at 9.6μm. Present understanding of atmospheric chemistry suggests that growth of atmospheric chlorofluorocarbons will eventually deplete stratospheric O_3 substantially and increase tropospheric O_3, with a net column-integrated reduction of O_3 (National Academy of Sciences, 1984). The net greenhouse effect of this scenario would be a slight warming, because of the greater greenhouse efficiency of tropospheric ozone molecules (Ramanathan and Dickinson, 1979; Lacis, 1984). In addition, increases of atmospheric methane and nitrous oxides both tend to increase tropospheric O_3 (Wuebbles et al., 1983; Cicerone, 1984), as do temperature effects caused by increasing CO_2. Recent measurements of O_3 are consistent with a long-term trend of increasing tropospheric ozone (Angell and Korshover, 1983), but the natural variability of O_3 prevents strong conclusions, especially regarding past trends. On the decadal time scale in the future, there is likely to be a significant, but quantitatively very uncertain, greenhouse warming due to O_3 changes.

Several conclusions follow from Figure 5.

- Non-CO_2 greenhouse gases are now adding to the greenhouse effect at a rate at least comparable to that of CO_2; this is a fundamental change from the situation in the period 1850 to 1970.
- The rate of increase of the greenhouse forcing is now three to 10 times greater than the mean rate for the period 1850 to 1958, caused by both increase in the growth rate of CO_2 and recent addition of other trace gases.
- A large degree of the greenhouse warming of gases added to the atmosphere in the past century has not yet occurred. This follows from the fact that much of these gases have been added in the past 25 years, a time comparable to or shorter than the thermal relaxation time of the ocean (Hansen et al., paper 2, 1984b).
- The greenhouse warming will soon rise above the level of natural climate variability, as shown quantitatively in the next section. This conclusion is independent of climate sensitivity for the range 2.5 to 5°C (4.5 to 9°F) for doubled CO_2, but the higher sensitivities imply that the eventual warming will be greater and that a larger proportion of the warming has not yet been realized.
- Large climate changes, with substantial impact on the earth's environment, will occur within the next several decades. This follows from the fact that effective doubling of CO_2, from, say, 300 to 600 ppm, requires CO_2 to increase only to 400 to 450 ppm because the other greenhouse gases will also be rising.

NEAR-TERM CLIMATE TRENDS

Experiments with a three-dimensional climate model have been initiated to estimate the expected transient climate response to a realistic rate of increase of greenhouse gases. For these experiments the climate model has an equilibrium global mean sensitivity of 4.2°C (7.6°F) for doubled CO_2. The ocean mixed-layer depth for this transient experiment was based on climatology, including geographical and seasonal variations. Heat capacity of the ocean beneath the mixed layer was included by treating heat perturbations as a passive tracer that could diffuse into the deeper ocean; the diffusion coefficent varied geographically as a function of the local climatological stability beneath the mixed layer.

The climate forcings included in these experiments are changes of greenhouse gases and stratospheric aerosols. Two cases, A and B, are believed likely to bracket the true rate of change of greenhouse forcing in the recent past and near future (say, 1958 to 1990). The same rates are extrapolated beyond 1990 in what is regarded as sensitivity experiments, since it is difficult to predict how the greenhouse forcing will change in the future. Case A includes reasonably well-documented changes of trace gases (CO_2, CCl_2F_2, CCl_3F, CH_4, and N_2O)* plus measured changes in stratospheric aerosol opacities (the two substantial aerosol events being Mt. Agung, 1963 to 1965, and El Chichon, 1982 to 1984), with a fixed mean aerosol opacity after 1984. Case B is similar to case A, but the chlorofluorocarbon increases are multiplied by 1.15 in an attempt to account for increases of other chlorine and fluorine compounds, and the CH_4 and N_2O growth rates are increased to account roughly for the O_3 and stratospheric H_2O changes suggested in Figure 5. The net greenhouse forcing is about 25% less in case A than in case B.

The computed global temperature trend is shown in Figure 6, along with observed temperatures through 1983. Although analysis of this experiment is beyond the scope of this paper, the following points bear mention.

There is strong month-to-month and year-to-year variability in both model and observations. Clearly in the model (and presumably in the real world) this is a natural variability, since the green-

---

*CO_2 increases as observed by Keeling for the interval 1958 to 1984 and with a slow growth rate into the future; ΔCO_2 is 15 ppm in the 1980s and 19 ppm in the 1990s. CH_4 increases by 0.5% per year in the 1960s, 1% per year in the 1970s, and 1.5% per year thereafter. N_2O increases by 0.2% per year in the 1970s and 0.3% per year thereafter. Future CCl_2F_2 and CCl_3F emission rates are taken as constant at the mean rate for the 1970s, with atmospheric lifetimes for the gases of 75 and 150 years, respectively; thus the annual increment in the atmospheric concentration of these gases decreases with time.

Figure 6

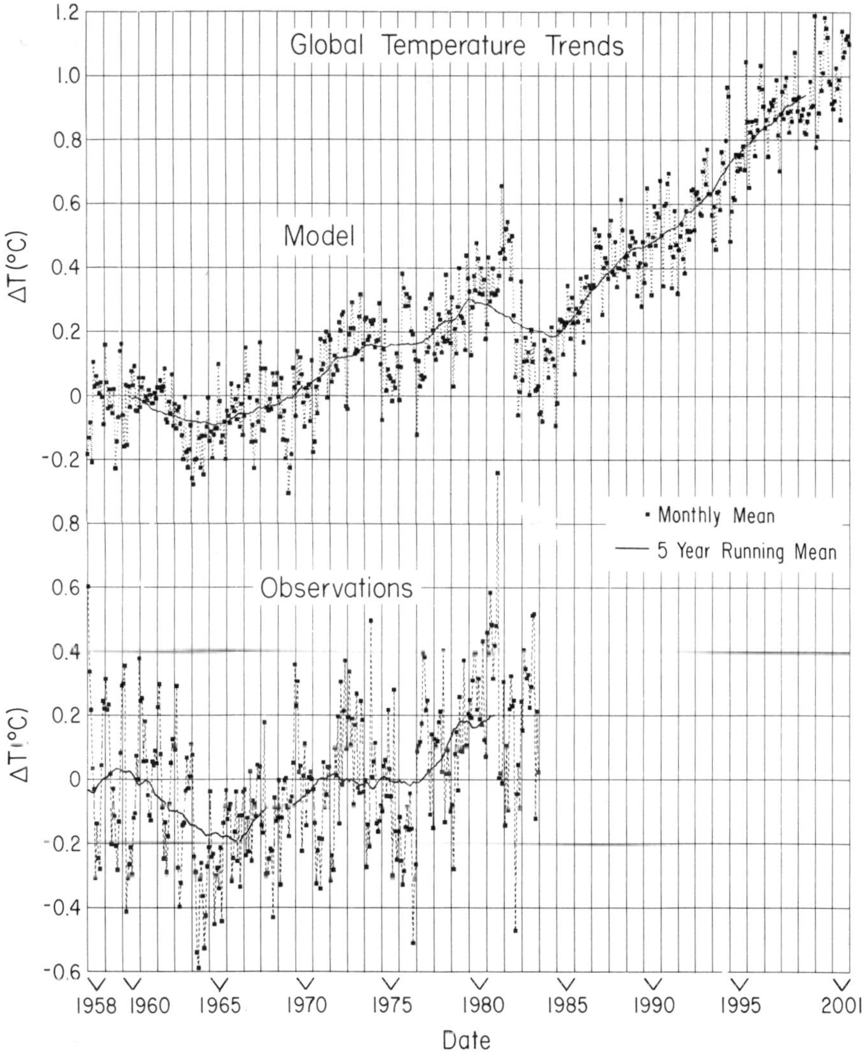

Modeled and observed global mean surface-air temperatures. The model climate is driven by Case B greenhouse gases and stratospheric aerosols as described in the text. The zero point for the model temperature is the mean for that month from the last 10 years of a 35-year control run with fixed January 1959 greenhouse gases and the mean 1880-1980 stratospheric aerosol optical depth. The observational data will be fully described elsewhere; the zero point for observations is the mean for that month from the period 1958 to 1983. Five-year running means are 61-month averages for the period from 30 months before to 30 months after the indicated month.

house forcing steadily increases throughout the period (with the exception of volcanic aerosols, which provide a large forcing only in the periods 1963 to 1965 and 1982 to 1984). The natural variability is less in the model than in the real world; this is expected since the model does not permit any variability in ocean dynamics.

The global warming in the model begins to rise above the level of natural variability by about 1990. It is about 0.5°C (0.9°F) warmer in 1990 than in 1970; by the year 2000 the warming has risen to about 1°C (1.8°F), which would bring global temperature to a level close to or exceeding the warmest period in the past 100,000 years (compare Figure 1 of Hansen et al., 1984a, based on original references in National Academy of Sciences, 1975).

In the model, a cooling of about 0.1°C (0.2°F) occurs after the Mt. Agung and El Chichon volcanoes *in the five-year smoothed temperatures*, with the minimum occurring about two years after the eruption. (The cooling in the model after El Chichon is greater, ~0.2 to 0.3°C [~0.4 to 0.6°F], if a shorter interval, one to three years, is used for smoothing the temperatures.) The model will need to be rerun a number of times through these time periods to verify the magnitude of the volcanic signal in the model. This cooling of about 0.1°C (0.2°F) in the model is smaller than that predicted by several other models (*compare* Robock, 1983), probably because of the greater effective ocean heat capacity in the Figure 6 model, which includes diffusive heat exchange beneath the mixed layer. The observations contain a suggestion of cooling after Mt. Agung erupted; the observations need to be extended before a useful search for an El Chichon signal can be made. (The data through 1983 suggest some cooling in 1982 and 1983, but it began several months before the El Chichon eruption. Interpretation is complicated by the presence of a major El Niño event in this period.)

For this model simulation, two cautions must accompany the results. First, although the model shows the temperature rising in the 1990s to levels well above those believed to have occurred in recorded history, this does not imply that temperatures are expected to increase at all times and places. For example, the natural variability of temperature becomes larger as smaller areas are considered. This is illustrated in Figures 7 and 8, which show the modeled and the observed temperatures for two areas of about 7×10^6 km^2 each; "western U.S." includes about two-thirds of the United States plus small areas of the Pacific Ocean and northern Mexico, while "eastern U.S." is about half ocean, including parts of the Gulf of Mexico and western Atlantic Ocean. For an area of this size, the natural variability of monthly mean temperatures

Figure 7

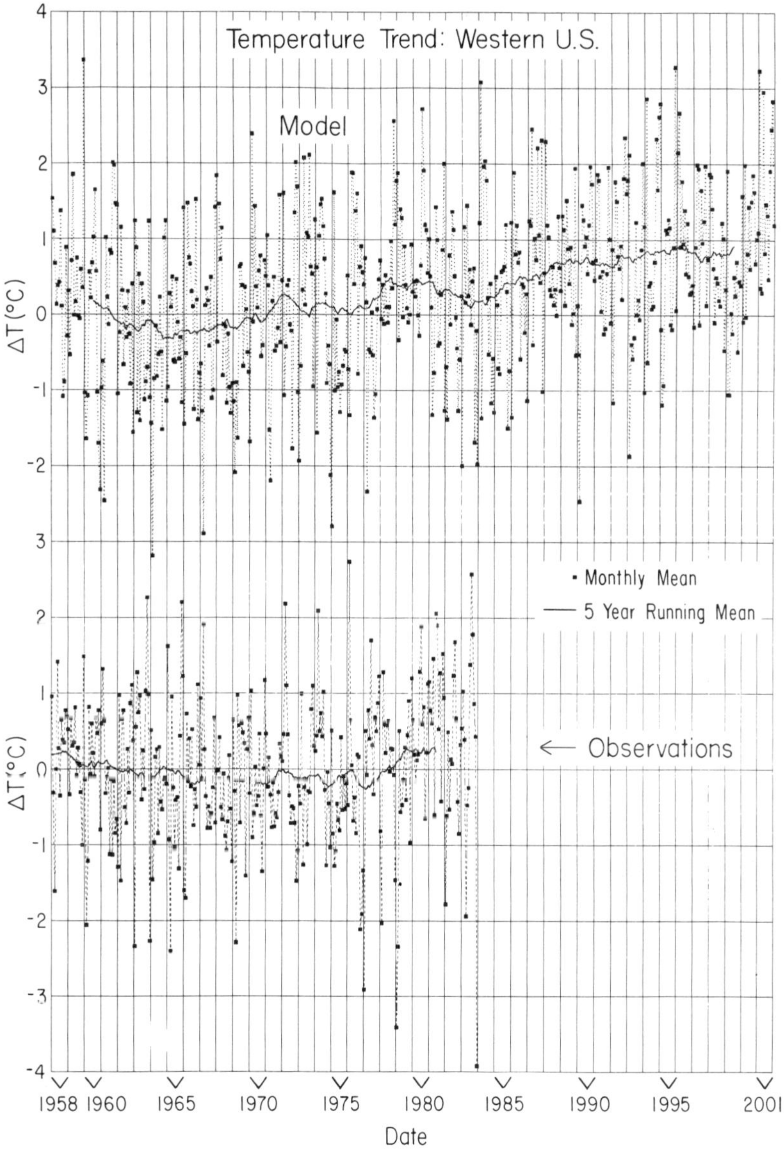

Same as Figure 6, but for longitude between 120°W and 90°W and latitudes between 23.5°N and 47°N.

Figure 8

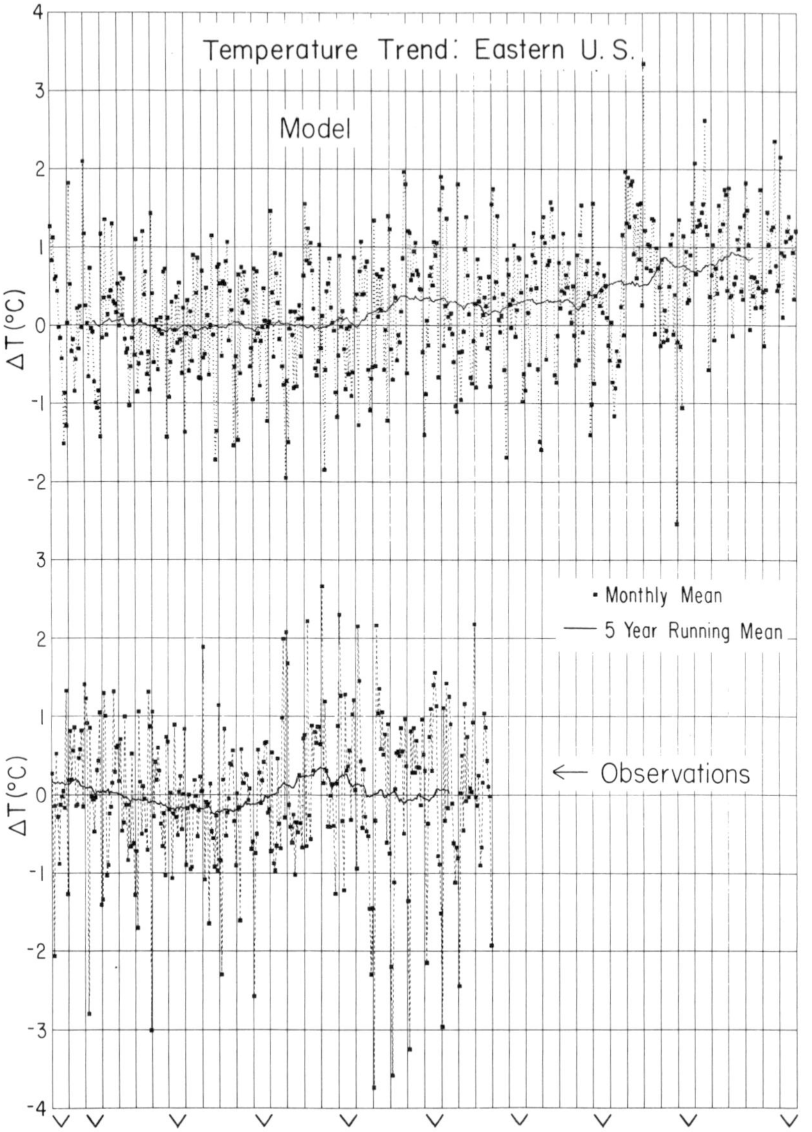

Same as Figure 6, but for longitudes between 90°W and 60°W and latitudes between 23.5°N and 47°N. Note that for this region (which is about 50% ocean) the natural variability of temperature in the model is less than for the "western U.S." (Figure 7), which is almost totally continental.

remains larger than the warming predicted in the next 15 years. Also, as a result of changes in atmospheric and ocean dynamics (the latter not allowed for in this current model), there may be some areas that would become cooler as the mean planetary temperatures rises. For example, if deep-water formation in the North Atlantic were greatly reduced, European temperatures could conceivably decrease (Bennett et al., 1984).

The second caution concerns the assumed climate forcings in the model and the many uncertainties in the model itself. Principal among the assumptions regarding climate forcings is the absence of any volcanic aerosol injections of Mt. Agung's or El Chichon's magnitude in the remainder of this century. If a large volcano erupts, the time at which the greenhouse warming rises above natural variability could be delayed several years. In addition, simulating the many physical processes in the climate model is uncertain; perhaps it is best to view simulations as sensitivity tests against which current and future climatic trends can be compared to obtain one overall model test. However, if the true climate sensitivity is in the range of 2.5 to 5.0°C (4.5 to 9°F) for doubled CO_2 (and if there is no large reduction in climate forcing, such as a large injection of volcanic aerosols), then, regardless of model details, global temperatures should begin to rise above the level of natural variability by the early 1990s.

CONCLUSIONS

Two National Academy of Sciences study groups (Charney, 1979; Smagorinsky, 1982) have estimated that a doubling of atmospheric CO_2 will cause a global mean warming of 3 ± 1.5°C (5.4 ± 2.7°F). Recent climate model results and empirical data present increasing evidence that their estimate of climate sensitivity is accurate. In addition, our climate model indicates that the climatic effects of greenhouse gases may begin to be apparent within the next decade, and that large climate changes are expected during the following decades.

The net effect of increased CO_2 on forests will conceivably be beneficial in the long term, in part because of "fertilization" by the added CO_2. But trees probably will require many decades to adjust their global distribution patterns to altered climate. Thus, during the next several decades there is also potential for detrimental effects, as climate patterns change and force tree populations to adapt. Because the time from tree planting to harvest is typically 30 to 100 years, forest managers should begin planning now to cope with climatic change.

REFERENCES

Angell, J. K. and J. Korshover. 1983. Global variation in total ozone and layer-mean ozone: an update through 1981. *J. Clim. Appl. Meteorol.* 22, 1611-1627.

Bennett, T., W. Broecker, and J. Hansen (ed.). 1984. *North Atlantic Deep Water Formation.* Report of a Workshop. Goddard Institute for Space Studies, New York.

Carbon Dioxide Assessment Committee. 1983. Changing Climate. National Academy Press, Washington, D.C.

Charney, J. (chairman). 1979. Carbon dioxide and climate: a scientific assessment. National Academy Press, Washington, D.C.

Cicerone, R. J. 1984. Methane in the atmosphere. *In* S. F. Singer (ed.), *Global Environmental Problems.* Paragon House, New York.

CLIMAP Project Members (A. McIntyre, project leader). 1981. Seasonal reconstruction of the earth's surface at the last glacial maximum. *Geological Society of America.* Map and Chart Series no. 36. Geological Society of America.

Craig, H., and C. C. Chou. 1982. Methane: the record in polar ice cores. *Geophys. Res. Lett.* 9, 1221-1224.

Ehhalt, D. C., R. J. Zander, and R. A. Lamontagne. 1983. On the temporal increase of tropospheric CH_4. *J. Geophys. Res.* ii, 8442-8446.

Hansen, J., A. Lacis, D. Rind, and G. Russell. 1984a. Climate sensitivity to increasing greenhouse gases, p. 57-77. *In Greenhouse Effect and Sea Level Rise.* Hutchinson Ross, Stroudsburg, Pa.

Hansen, J., A. Lacis, D. Rind, G. Russell, P. Stone, I. Fung, R. Ruedy, and J. Lerner. 1984b. Climate sensitivity: analysis of feedback mechanisms. *In* J. E. Hansen and T. Takahashi (ed.), *Climate Processes and Climate Sensitivity.* American Geophysical Union, Washington, D.C.

Hansen, J., G. Russell, D. Rind, P. Stone, A. Lacis, S. Lebedeff, R. Ruedy, and L. Travis. 1983. Efficient three-dimensional global models for climate studies: models I and II. *Mon. Weather Rev.* 111, 609-662.

Lacis, A. 1984. Chlorofluorocarbons and stratospheric ozone. *In* S. F. Singer (ed.), *Global Environmental Problems.* Paragon House, New York.

Manabe, S., and R. J. Stouffer. 1980. Sensitivity of a global climate model to an increase of CO_2 concentration in the atmosphere. *J. Geophys. Res.* 85, 5529-5554.

Milankovitch, M. 1941. Kanon der Erdbestrahlung und seine Andwendung auf das Eiszeitenproblem. *R. Serv. Acad. Spec. Publ.* 133, 1-633.

National Academy of Sciences. 1975. *Understanding Climate Change*. National Academy Press, Washington, D.C.

National Academy of Sciences. 1984. *Causes and Effects of Changes in Stratospheric Ozone: Update 1983*. National Academy Press, Washington, D.C.

Oeschger, H. 1984. Long-term climate stability. Thirteenth International Conference on the Unity of the Sciences, Sept. 2-5, 1984, Washington, D.C.

Ramanathan, V., and R. E. Dickinson. 1979. The role of stratospheric ozone in the zonal and seasonal radiative energy balance of the earth-troposphere system. *J. Atmos. Sci.* 36, 1084-1104.

Robock, A. 1983. The dust cloud of the century. *Nature* 361, 373-374.

Schiffer, R. A., and W. B. Rossow. 1983. The International Satellite Cloud Climatology Project (ISCCP): the first project of the World Climate Research Programme. *Bull. Am. Meteorol. Soc.* 64, 779-784.

Schneider, S. H. and R. E. Dickinson. 1974. Climate modeling. *Rev. Geophys. Space Phys.* 12, 447-493.

Smagorinsky, J. (chairman). 1982. Carbon dioxide and climate: a second assessment. National Academy Press, Washington, D. C.

Washington, W. M. and G. A. Meehl. 1984. Seasonal cycle experiment on the climate sensitivity due to a doubling of CO_2 with an atmosphere general circulation model coupled to a simple mixed layer ocean model. *J. Geophys. Res.* (submitted).

Weiss, R. F. 1981. The temporal and spatial distribution of tropospheric nitrous oxide. *J. Geophys. Res.* 86, 7185-7195.

World Meteorological Organization. 1983. World Meteorological Organization project on research and monitoring of atmospheric CO_2. Report no. 10, R.J. Bojkov, ed. World Meteorological Organization, Geneva.

Wuebbles, D. J., F. M. Luther, and J. E. Penner. 1983. Effect of coupled anthropogenic perturbations on stratospheric ozone. *J. Geophys. Res.* 88, 1444-1456.

PREDICTING REGIONAL CLIMATE CHANGE: IMPROVING THE MODELS

by David Rind

ABSTRACT

With the increasing atmospheric concentration of trace gases such as carbon dioxide (CO_2), the question of whether regional climate will change becomes increasingly important. The likelihood of such changes, in the hydrological cycle, is explored in two ways, first by looking at indicators of past climate changes, and then by reviewing results from general circulation models run with elevated atmospheric CO_2 levels. The probability of substantial climate change on the regional level exists, although models are currently too crude to predict the nature of these changes with great certainty.

Climate is defined as the time-averaged value of meteorological quantities. Implied in the concept is that the averaging is taken over a long enough time to remove any inherent variability. This is especially important when discussing "regional" climate. In practice, the values reported as climatological averages represent anywhere between 30 and 100 years of data. There is no guarantee that such a data set represents the true long-term average, or even that there really is a meaningful long-term average. In fact, it is now apparent that "climate" actually refers to a nonstationary process--the time-averaged weather does not stay the same. Changes in climate are caused, potentially, by changes in solar radiation, changes in atmospheric composition, or changes in the solid/liquid geophysical system (topography or ocean circulation). The variability occurs on all time scales and has implications for all spatial scales, including the regional. The importance of the variability depends on the process being considered.

It is thought, for example, that the difference between the "climatic optimum," or warmest time period, some 6000 and 8000 years ago, and the coldest period, some 18,000 years ago, amounts to about 5°C (9°F) on the global average (Figure 1, top). Yet 18,000 years ago, glaciers covered much of the North American continent, an occurrence that obviously had regional implications in the areas directly affected. However, even areas south of the ice sheet experienced significant changes.

Figure 1 (middle) shows the fluctuations in lake levels in the southwestern United States. During the last ice age, lake levels were high. Although it is uncertain how precipitation changed, temperatures were colder and evaporation should have been reduced. During the climatic optimum, lake levels were at their lowest. This picture clearly emphasizes that over hundreds and thousands of years, the hydrological aspect of climate undergoes large fluctuations even on a regional level. Figure 2 offers a global perspective for changes between cold and warm periods. In direct opposition to the variation over the southwestern United States, African lake levels were high during the warm period and low at the last glacial maximum, emphasizing that the type of variation experienced in particular regions can be anticorrelated. Analysis needs to be done on a regional basis.

The prevailing theory, based on the cyclic behavior of the changes, is that orbital variation of the earth around the sun is responsible for long-period climate fluctuations. Although such variations do not produce large alterations of the total solar radiation received on the planet, they do change the seasonal and latitudinal distribution and perhaps trigger feedbacks such as CO_2 concentrations (Figure 1, bottom). Clearly, such large-amplitude changes would be important to every community. What about shorter, more moderate variations? Figure 3 (top) shows the change in globally averaged temperature during the past century. The total range amounts to about 0.8°C (1.4°F). A comparison between a warm period (1940 to 1960) and a cold period (1900 to 1920) shows that precipitation was greater in the eastern United States and less in the Southwest during the warm period (Figure 3, middle). The changes in the centers of these regions of increase and decrease amount to about 10% of the regions' annual average rainfall. The decrease in precipitation in the Southwest during the warmer decades is consistent with the decrease in lake levels during the climatic optimum, another example of significant regional variation in alterations of the hydrological cycle.

What caused the temperature change during the past century? Although some of it undoubtedly represents natural internal variability, the effect of increased atmospheric carbon dioxide (CO_2) may also be evident. Figure 3 (bottom) shows the carbon dioxide level as measured at Mauna Loa for the past 25 years. The annual

Figure 1

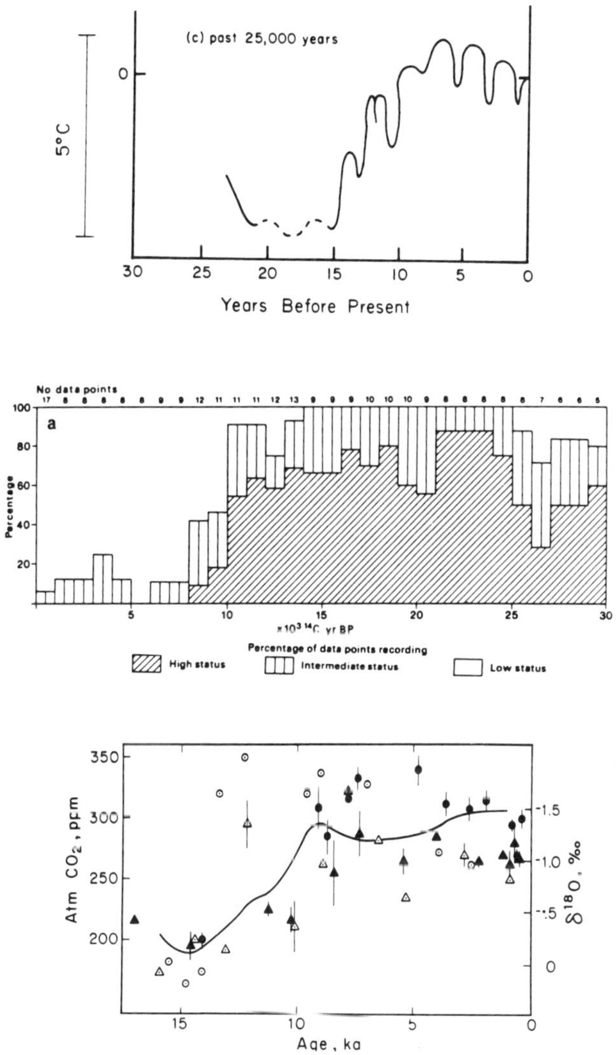

Top: Global temperature trend for the past 25,000 years based on changes in tree lines, fluctuations of alpine and continental glaciers, and shifts in vegetation patterns recorded in pollen spectra (National Academy of Sciences, 1975), with the temperature scale set by three-dimensional climate model simulation for 18,000 years ago (Hansen et al., 1984a). Middle: Histogram of lake-level status for 1000-year time periods from 30,000 years B.P. (before present) to present for the southwestern United States (Street and Grove, 1979). Bottom: Ice-core evidence for change in atmospheric CO_2 during deglaciation. Circles: Greenland cores. Triangles: Antarctic cores. (From Berger and Keir, 1984.)

Figure 2

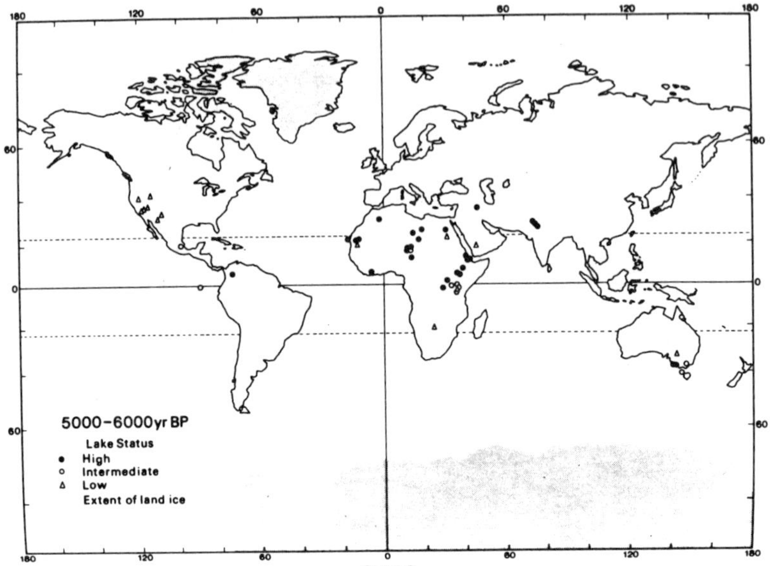

FIGURE 9

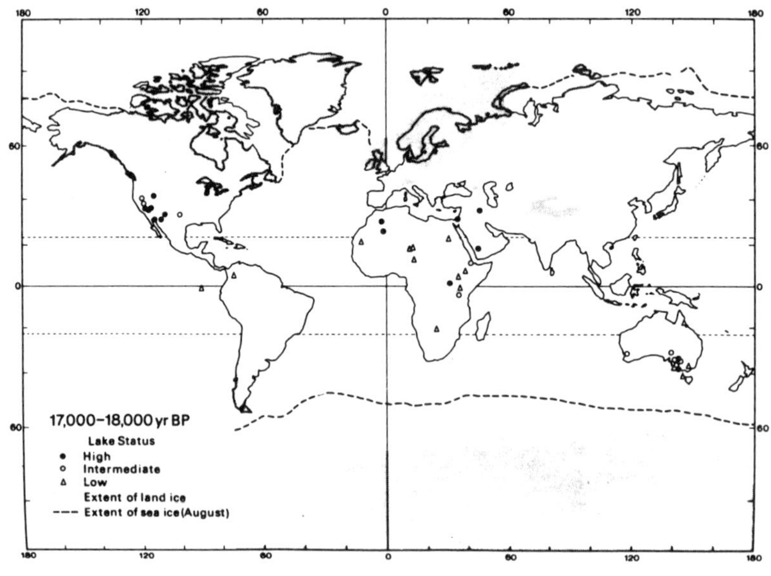

Lake level status at 5000 to 6000 years B.P. (top) and 17,000 to 18,000 years B.P. (bottom) (From Street and Grove, 1979.)

Figure 3

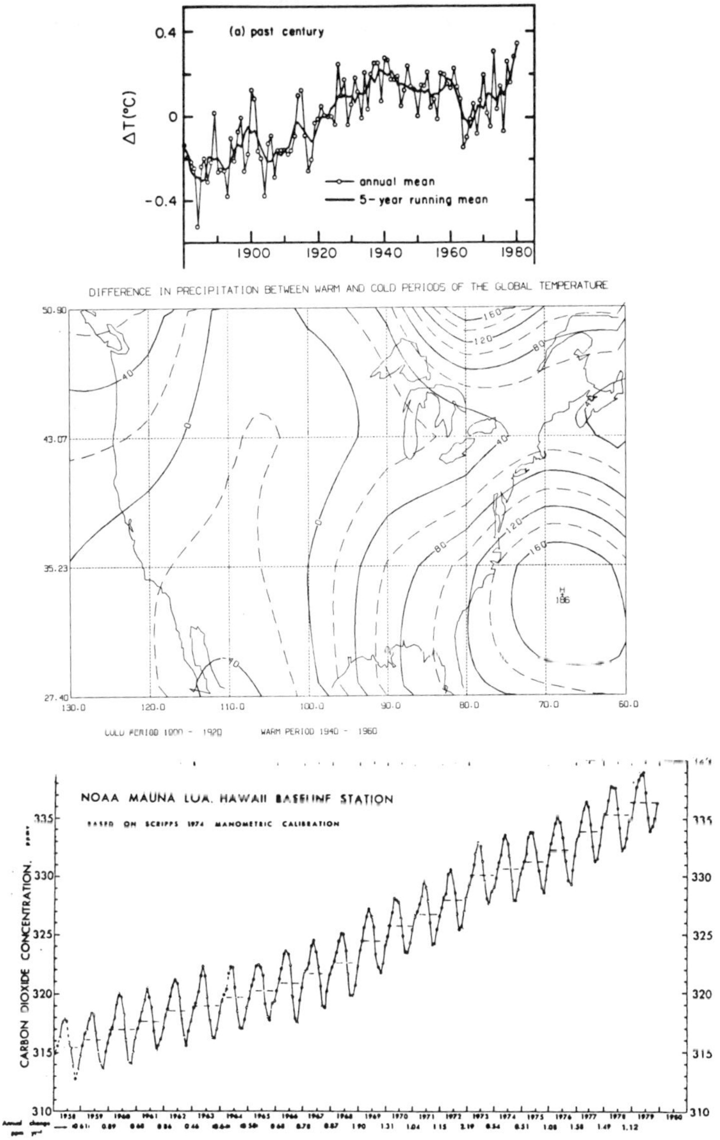

Top: Global temperature trend for the past century (Hansen et al., 1984a). Middle: Precipitation difference (millimeters per year) between warm period (1940-1960) and cold period (1900-1920). (From Lebedeff, 1986.) Bottom: CO_2 level recorded at Mauna Loa, Hawaii in parts per million and annual changes. (Reproduced courtesy of National Oceanic and Atmospheric Administration.)

change per year in atmospheric concentration is about 1 part per million (ppm) (0.3%). It is estimated that the change since 1880 amounts to about 70 ppm, which would account for a global temperature increase of about 0.5°C (0.9°F), consistent with the magnitude of change thought to have occurred. As CO_2 concentrations continue to increase and global temperature rises, impacts on regional hydrological cycles are likely.

MODEL PREDICTORS

In two reports, the National Academy of Sciences (1979, 1981) estimated that the coming doubling of atmospheric carbon dioxide would raise the global mean temperature 3 ± 1.5°C (5.5 ± 2.7°F). This magnitude of warming, which would probably not be fully realized before the end of the next century, is much larger than the change experienced within the last century and is of the same order as the change from glacial to interglacial ages, which normally takes thousands of years. Lacis et al. (1981) have shown that the climatic impact of other anthropogenic releases during the 1970s is comparable to that of increasing CO_2, and the rising levels of these other trace gases will likely increase the rapidity with which the global warming occurs. What will the regional effects be of this coming climate change, and how are they to be predicted? Past climates cannot be used as analogues, because the climatic record shows (e.g., Figure 1, top) that temperatures warmer than those of the present by 3°C (5°F) have not occurred for many thousands (if not millions) of years. Further, the climate changes associated with increasing amounts of greenhouse gases may not be similar to those of the past, which probably were associated with different forcings.

The National Academy's reports were based largely on the result of predictions made with general circulation models (GCMs). These are numerical models of the earth-atmosphere system that solve the basic equations for atmospheric motion and provide realistic boundary conditions of the ground and ocean (e.g., Hansen et al., 1983). The models are first run with the current atmospheric conditions and simulate the current climate reasonably well. They are then used with doubled atmospheric carbon dioxide to simulate a new climate. The difference between these two equilibrium climate simulations provides an estimate for how the climate system will change.

The models need to include parameterizations for many aspects of the physical system, and some of these are extremely crude. There are only a few such models, and each modeling group has formulated its model somewhat differently. The result is that the models are not in perfect agreement as to their prediction of the magnitude of the coming warming (which is the reason for the range in the National Academy's estimate). For example, the model developed at Geophysical Fluid Dynamics Laboratory, Princeton, predicts a warming of about 2°C (4°F) for doubled CO_2, while the models developed at the National Center for Atmospheric Research (NCAR) and Goddard Institute for Space Studies give values close to 4°C (7°F). One difference between these models relates to the prediction of cloud-cover change. The GFDL model does not predict clouds; Hansen et al. (1984b) have shown that cloud cover is the primary reason for the different model sensitivities.

The models predict all of the climate variables at all locations on the earth, with specific horizontal resolutions. In particular, they show changes in the hydrological cycle, with an overall intensification of evaporation and precipitation in the warmer climate regime. However, the regional variations are large, and the different models vary in their predictions of what changes will occur from place to place. Figure 4 shows the change in soil moisture (or precipitation minus evaporation, which is closely related) for different model results. The top two panels in Figure 4 show the soil-moisture change in the GFDL model for quadrupled CO_2 (which gives the same magnitude warming as the other models). Two paired model results are shown, with the lower panel from a model with finer horizontal resolution. Both paired top panels in Figure 4 show a drying (decrease in soil moisture) throughout the United States, although the patterns differ with resolution.

The two middle panels in Figure 4 are from two experiments at GISS, which differ in the physics of sea-ice formation. The experiment providing results shown in the left-hand panel of these two has a global warming caused by doubled CO_2 of 4.3°C (7.7°F), while the other experiment gives a warming of 4.8°C (8.6°F). With the more modest warming, drying is predicted over the United States in the Southwest and Northeast; in the other experiment it is drier in the region to the west of the Great Lakes.

The bottom panel of Figure 4 shows the soil-moisture change in the NCAR experiment. The United States is drier over the Southeast and from the north-central portion of the country into southern Canada, while it is wetter in the Northeast. These results indicate that, with the current state of modeling capability, results depend on model resolution and model physics and vary from one model to the other.

Figure 4

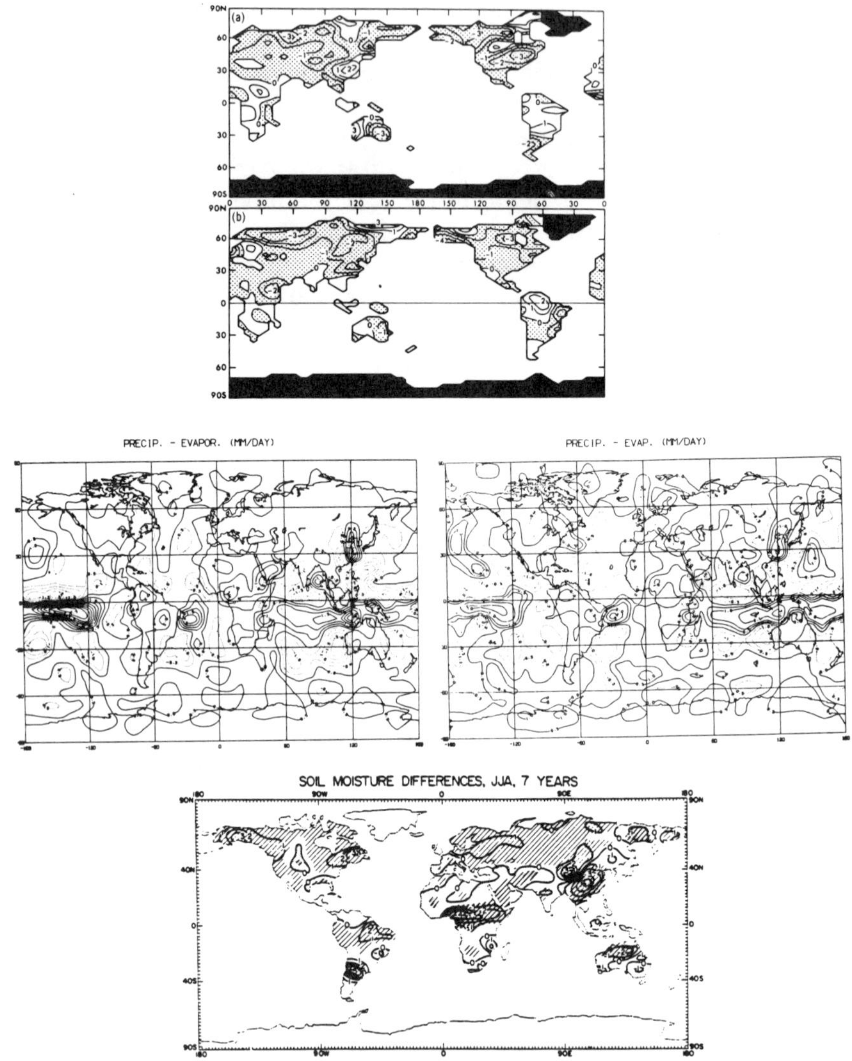

Top: Difference in soil moisture (centimeters) in the GFDL model for a quadrupled-CO_2 simulation for June through August for a 15-wave spectral model (top) and a 21-wave (increased-resolution) spectral model (bottom). (From Manabe et al., 1981.) Middle: Difference in precipitation minus evaporation (millimeters per day) in the GISS model for a doubled-CO_2 simulation for June through August for models with global warming of 4.3°C (7.7°F) (left) and 4.8°C (8.6°F) (right). Bottom: Difference in soil moisture (centimeters) in the NCAR model for a doubled-CO_2 simulation for June through August (Washington and Meehl, 1984).

MODEL OUTPUT

As the models solve for all the climate variables (winds, temperature, rainfall, evaporation, cloud cover, etc.), they are capable of producing a mass of diagnostics of potential use to various communities. The results in this section come from the GISS model with 4.3°C (7.7°F) warming caused by doubled CO_2.

Figure 5 presents the change in growing season over the United States in the doubled-CO_2 climate as simulated by the model. Throughout most of the northern United States and Canada, the growing season increases by 20% or more as the climate has warmed. Again there is regional variability, depending on the values for the current climate and the precise location. Figure 6 shows the change in several parameters for the grid box covering the northwestern portion of the United States as a function of month. Except for the autumn months, there is a substantial increase in the soil moisture (as can be seen in Figure 4, this area experiences an increase in precipitation minus evaporation). Snow cover, in general, decreases. The plant water stress (see Rind and Lebedeff, 1984, for the exact definition) increases in summer with the warmer weather. A drought index (defined in Rind and Lebedeff, 1984) indicates the degree of drought represented by successive months of below- or above-normal precipitation-minus-evaporation conditions.

This grid box experiences reduction in drought conditions throughout the year. The results for the Northwest grid box, for a grid box including the Colorado River, for the Southern Plains, and for the northeast United States, respectively, are shown in Table 1. The Northwest is much wetter, the Southern Plains somewhat drier, and the Northeast considerably drier. Obviously these results cannot be thought of as definitive, given the current state of modeling, but they do represent the types of output the models produce.

NECESSARY MODEL IMPROVEMENTS

As indicated by the preceding discussion, both physical understanding and modeling capability need improvement before model predictions can become reliable. Several areas need particular improvement.

- The atmosphere interacts continually with the ocean, and changes in the ocean's circulation or ability to absorb heat and carbon dioxide will alter the nature, magnitude, and timing of the coming climatic change. Ocean models are far

Figure 5

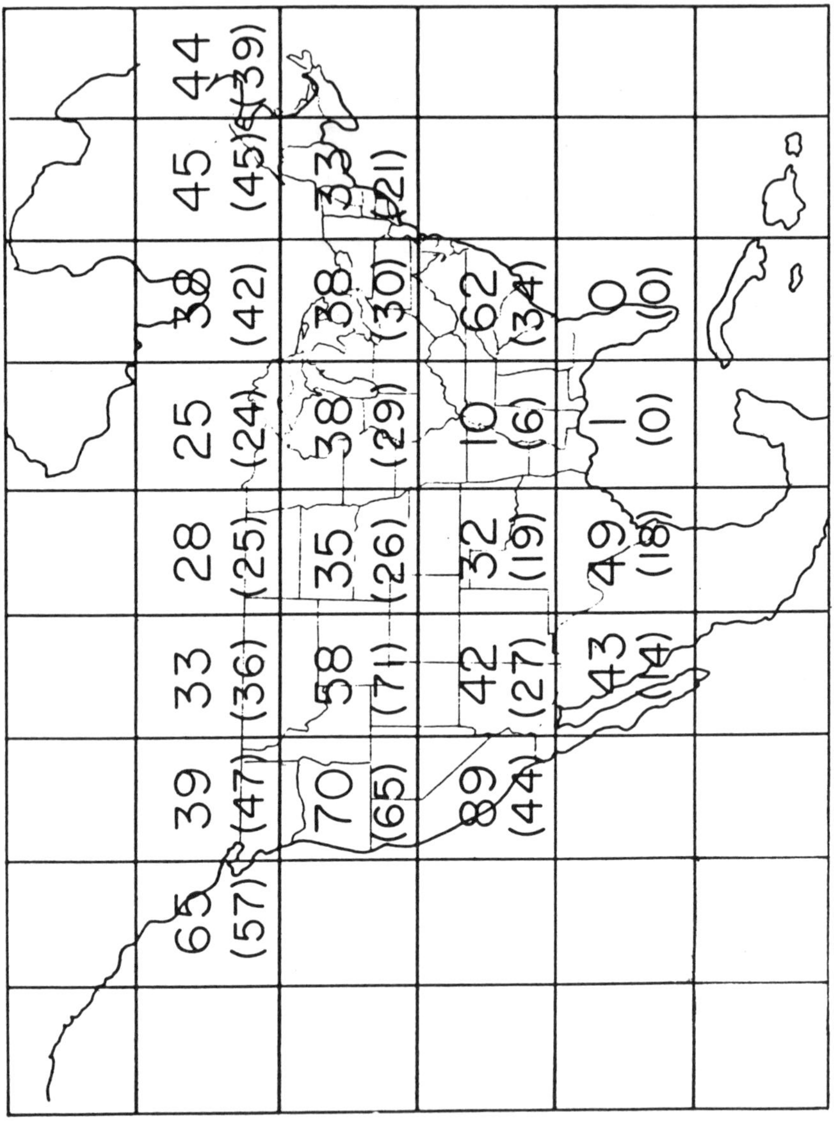

Change in the growing season in the GISS CO_2 simulation. The top number indicates the actual change (days); the bottom number (in parentheses) gives the change in percent relative to the control run (Rind and Lebedeff, 1984). The growing season is defined as the time between the last spring frost and the first autumn frost. The grid box size defines the horizontal resolution of the model.

Figure 6

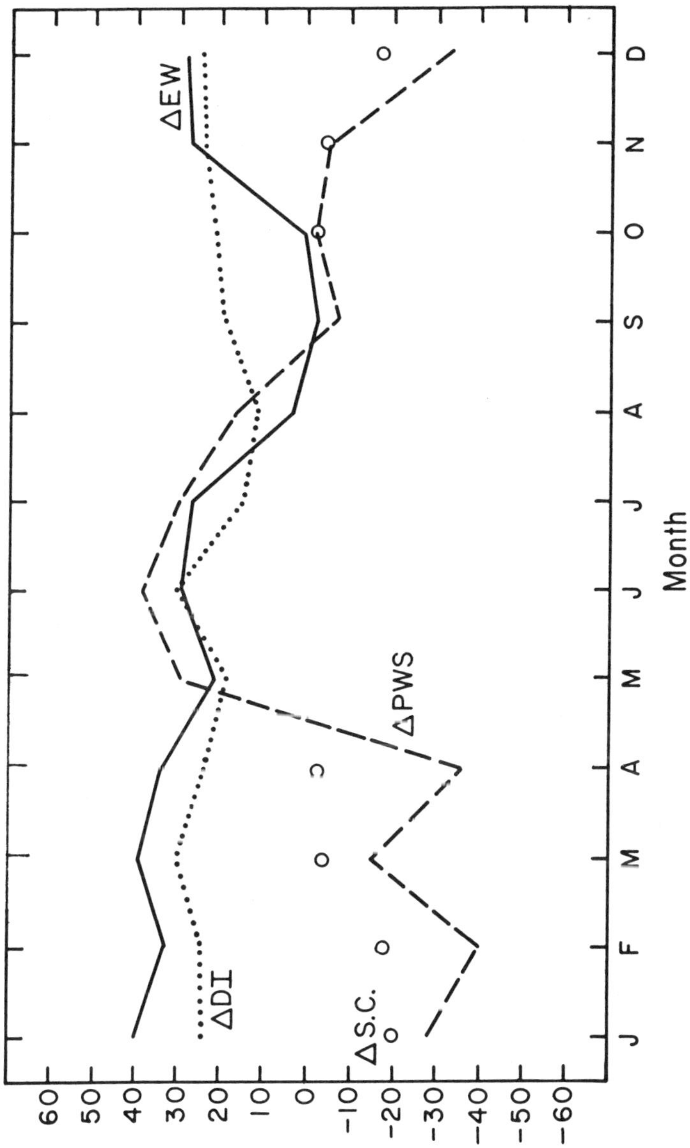

Change in total earth water (EW) (millimeters), drought index (DI) (times 10), plant water stress (PWS), and snow cover (S.C.) (percent) in the GISS doubled-CO_2 simulation for the grid box encompassing the northwestern United States (Rind and Lebedeff, 1984).

Table 1

	Grid box 9		Grid box 16		Grid box 17		Grid box 14	
	Control	2'd CO_2	Control	2'd CO_2	Control	2'd CO_2	Control	2'd CO_2
Extremely dry	4	2	4	1	0	0	0	22
Severely dry	8	4	4	8	0	2	3	13
Moderately dry	13	9	13	12	3	16	13	12
Mildly dry	13	11	13	17	21	18	10	12
Near normal	38	16	26	23	37	40	33	20
Mildly wet	8	9	19	19	18	12	18	18
Moderately wet	8	11	8	10	7	5	6	0
Severely wet	7	10	8	12	11	5	10	2
Extremely wet	1	27	5	0	3	2	7	1

Drought conditions for the control and doubled CO_2 (2'd CO_2) experiment for the northwestern United States (grid box 9), Colorado River basin (grid box 16), Southern Great Plains (grid box 17), and northeastern United States (grid box 14) (Rind and Lebedeff, 1984). The columns indicate the percentage of times the drought index indicated specific conditions throughout the year.

more primitive than atmospheric models, due partially to the difficulty in obtaining ocean observations. However, the ocean's role in impacting climatic change needs to be considered.

- Cloud cover can amplify or diminish the magnitude of climatic change and will affect specific latitudes and regions. Therefore, the ability to model clouds for the current climate must be upgraded before the predicted cloud-cover changes can be believed.
- The local ground hydrology, the movement of water into and out of the soil/vegetation system, can have a large impact on the hydrological cycle change regionally. This cycle involves complicated processes which scientists have only recently begun to investigate from a global modeling perspective.
- The necessary horizontal resolution in models needs to be investigated; as the horizontal resolution increases, the computing time increases substantially, which means most models are, of necessity, run with coarse resolution. The ready availability of computer resources would alleviate this problem.
- The sophistication with which the earth-atmosphere system is modeled needs to be improved before the model predictions of specific regional changes can be accepted.

CONCLUSIONS

Paleoclimate evidence, both long- and short-time scale, with large and small magnitude, indicates that regional hydrological changes will likely accompany global temperature changes. GCMs are being used to estimate what climate changes will occur in a world with carbon dioxide building up; however, the current state of modeling is not sufficiently honed to predict effects of rising CO_2 levels on particular, discrete regions of the globe. GCMs can produce information on climate variables potentially applicable to regional concerns--e.g., effects on specific forest species--and greater interaction between communities that could make use of such results and the modelers is desirable, especially as models improve.

REFERENCES

Berger, W. H., and R. S. Keir. 1984. Glacial-Holocene changes in atmospheric CO_2 and the deep-sea record. *In* J. E. Hansen and T. Takahashi (ed.), *Climate Processes and Climate Sensitivity*. AGU

Monograph 29. American Geophysical Union, Washington, D.C.

Hansen, J., G. Russell, D. Rind, P. Stone, A. Lacis, S. Lebedeff, R. Reudy, and L. Travis. 1983. Efficient three-dimensional global models for climate studies: models I and II. *Mon. Wea. Rev.* 111, 609-662.

Hansen, J., A. Lacis, D. Rind, and G. Russell. 1984a. Climate sensitivity to increasing greenhouse gases. *In Sea Level Rise to the Year 2100.* Van Nostrand, New York.

Hansen, J., A. Lacis, D. Rind, G. Russell, P. Stone, I. Fung, R. Reudy, and J. Lerner. 1984b. Climate sensitivity: analysis of feedback mechanisms. *In* J. E. Hansen and T. Takahashi (ed.), *Climate Processes and Climate Sensitivity.* AGU Monograph 29. American Geophysical Union, Washington, D.C.

Lacis, A., J. Hansen, P. Lee, T. Mitchell, and S. Lebedeff. 1981. Greenhouse effect of trace gases, 1970-1980. *Geophys. Res. Lett.* 8, 1035-1038.

Lebedeff, S. 1986. On the possible effects of the global warming on the precipitation pattern over North America. *Mon. Wea. Rev.* (submitted).

Manabe, S., R. T. Wetherald, and R. J. Stouffer. 1981. Summer dryness due to an increase of atmospheric CO_2 concentration. *Climatic Change* 3, 347-386.

National Academy of Sciences. 1975. *Understanding Climatic Change.* National Academy Press, Washington, D.C.

National Academy of Sciences. 1979, 1981. *Carbon Dioxide and Climate, a Scientific Assessment.* National Academy Press, Washington, D.C.

Rind, D., and S. Lebedeff. 1984. Potential climatic impacts of increasing atmospheric CO_2 with emphasis on water availability and hydrology in the United States. Environmental Protection Agency, Washington, D.C.

Street, F. A. and A. T. Grove. 1979. Global maps of lake-level fluctuations since 30,000 yr B.P. *Quaternary Res.* 12, 83-118.

Washington, W., and G. Meehl. 1984. Seasonal cycle experiment on the climate sensitivity due to a doubling of CO_2 with an atmospheric general circulation model coupled to a simplified mixed layer ocean model. NCAR/8041/82-1(E). National Center for Atmospheric Research, Boulder, Colo.

AN APPROACH FOR GENERATING CLIMATE CHANGE HYPOTHETICALS GIVEN LIMITATIONS IN CURRENT CLIMATE MODELS

by Michael J. Gibbs and John S. Hoffman

ABSTRACT

This paper discusses (1) the need for hypothetical climate-change scenarios that describe the ways in which climate may change as a result of a greenhouse warming and (2) a method for creating such scenarios using outputs from current climate models. The need to make preliminary estimates of the likely impacts of the impending climate change drives the desire for geographically- and temporally-resolved estimates of climate change. Such estimates, in turn, will allow a determination to be made of the value of investing additional resources in scientific research to improve our ability to forecast regional climate change over time. A method for developing internally-consistent hypothetical scenarios is proposed and examples of the estimates obtained are presented.

THE NEED FOR CLIMATE CHANGE SCENARIOS

The world's climate is undergoing an unprecedented man-induced change. Atmospheric concentrations of "greenhouse gases," which include carbon dioxide (CO_2), methane, nitrous oxide, and chlorofluorocarbons, have been increasing over the past century and are expected to continue to increase in the foreseeable future. Two National Academy of Sciences panels have concluded that the projected increases in the atmospheric concentrations of these gases will almost certainly produce a significant global warming.

Climatologists and climate modelers are working to project future climate change not only for the the entire globe, but also on a regional scale. The primary tools for these investigations are general circulation models, or GCMs. These models use simplified representations of the physics and dynamics of the global climate system to simulate daily weather. By varying the specifications of the atmospheric constituents in the models, it is possible to assess how global climate may change as a result of the addition of greenhouse gases to the earth's atmosphere. For example, by specifying the earth's current atmosphere, one set of climate statistics can be simulated. Then, using an alternative atmosphere (with higher concentrations of greenhouse gases), a second set of climate statistics can be estimated and compared with the first set. The resulting comparison is an estimate of climate change.

Unfortunately, GCMs are not yet able to produce the forecasts of climate change needed by business and environmental planners. For example, the projections from current GCMs cannot provide information about climate change at a level of geographic resolution that is refined enough for planning purposes. In addition, because GCMs require so much computer time, most computer experiments have not simulated enough years of weather to estimate the interannual variability that can be expected with a changed atmosphere. Finally, to date, only one GCM has been used to estimate changes in climate as the atmosphere changes over time. Work has generally focused on simulating climate with a future atmosphere in which CO_2 has been assumed to have doubled and all other factors (such as oceans) have come into equilibrium.

To forecast climate reliably, considerable resources will be necessary to collect data and improve the models. The time required for these tasks will depend on the level of resources invested in scientific data collection and research over the next decade. Current investments are almost certainly insufficient to produce a climate forecasting capability within the next decade.

Increased investment in scientific research cannot be made without careful consideration of the potential utility of climate forecasts for environmental and economic decision-making, within both the government and the private sector. Unfortunately, current models cannot provide estimates of the speed and location of possible climate changes, which are necessary for determining the value of climate forecasts. To surmount this dilemma, we must develop hypothetical scenarios that provide a rough first approximation of climate change, which can then be used to assess its impacts. To accomplish this, the results of current GCMs must be translated into more usable forms. The approach presented here is intended as a step in that direction.

It must be emphasized that climate change scenarios will not, in general, be useful for planning purposes. Their primary utility will be in providing a basis for performing preliminary impact assessments; the results of these assessments can help determine the importance of better forecasts of climate change. If these analyses show that climate change would have large economic or environmental impacts that could affect many decisions, a basis may exist for accelerating scientific efforts aimed at estimating future climate change reliably enough for planning.

In a very few cases, hypothetical scenarios may allow some users to evaluate the risks associated with current decisions. Although individual scenarios will not reliably describe the future for any given area, the range of change implied by a variety of scenarios may provide a better notion of future climate than would statistics based on extrapolations of past climate. Therefore, the scenarios may provide a basis upon which to assess the vulnerability of certain climate-sensitive decisions.

The remainder of this paper describes the following:

- the desired characteristics of hypothetical climate change scenarios;
- the general circulation models available to form the bases for the scenarios;
- the major assumptions required to use currently-available data to create scenarios;
- sample results for one scenario which describes several small areas; and
- the improvements needed in the methodology.

We conclude with an appendix that provides a step-by-step description of the method used to generate the sample results.

CHARACTERISTICS OF HYPOTHETICAL CLIMATE CHANGE SCENARIOS

To be most useful in performing preliminary impact analyses, the scenarios should have the following characteristics:

- they should incorporate and reflect the full range of current estimates of potential climate change;
- they should be objective and internally consistent; and
- they should describe climate change in quantities most useful for researchers and planners.

To reflect the full range of climate change estimates, the results of all the major GCMs must be compared and contrasted. Because GCMs cannot yet provide reliable estimates of climate change at the regional level, there is considerable uncertainty regarding potential climate changes for small areas. This uncertainty can be reflected, in part, by using the full range of estimates provided by existing GCMs and creating a series of scenarios from them.

To make the scenarios as objective and internally consistent as possible, the various assumptions and techniques used to create them (discussed below in the section titled Model Assumptions Required) must be examined and verified. Because there is uncertainty about some of the required assumptions, a range of assumptions may be necessary to reflect the full range of potential climate change. Care must be taken to vary these assumptions in ways which are internally consistent. In this way, the scenarios would reflect the uncertainty surrounding these assumptions as well as the uncertainty inherent in the GCMs.

Finally, the scenarios must describe climate change in terms that are useful for research. In general, this means providing estimates of changes for small areas over time. Quantities of interest may include:

- change in mean monthly variables such as temperature, precipitation, ground wetness, and runoff;
- changes in the expected frequency of specific events such as droughts of given severities and storms of given intensity and duration;
- changes in the frequency of specific events, estimated by simulating future weather and climate using probabilistic and statistical techniques; and
- changes in the expected ranges and distributions of variables such as monthly temperature and precipitation.

In this first attempt at creating hypothetical climate change scenarios, we used a single GCM,[1] one set of assumptions, and only computed estimates of changes in mean monthly temperature and precipitation. To develop a full set of scenarios with the above characteristics, additional work is required to compare and contrast the various GCM results, develop alternative sets of assumptions, and compute changes in other climate variables of interest.[2] The next section discusses the strengths and weaknesses of current GCMs.

---

[1]The output of the Manabe/Wetherald model at the National Oceanic and Atmospheric Administration's Geophysical Fluid Dynamics Laboratory (GFDL) was used in this project. The gracious cooperation of GFDL in supplying these data does not imply an endorsement of our method of creating scenarios.

[2]This work is ongoing. The results of three GCMs are being used to develop more complete scenarios than those presented here. The GCMs being used were developed at the Geophysical Fluid Dynamics Laboratory, Princeton, N.J., the Goddard Institute for Space Studies, New York, N.Y., and the National Center for Atmospheric Research, Boulder, Colo.

GCMs

Several full-scale GCMs exist in the United States. Each divides up the earth into a series of regions called grid boxes. Each grid box is comprised of a series of layers which represent land, ocean, and layers of the atmosphere. Thus, these models consist of numerous three-dimensional "boxes." Global weather is simulated using difference equations which describe the physics and dynamics of how energy and material move among the boxes. Statistics are computed for a variety of quantities that describe the simulated weather conditions in the boxes (Table 1).

The boxes in the models are defined as simplified representations of the earth's characteristics. For example:

- Geographic features of the earth's surface are represented by averaging the topographic features within each grid box.
- Oceans are represented as heat sinks, with a mixed layer on top of a colder deep layer. Some models do not circulate the waters of the ocean. Others use current ocean circulation patterns but do not change the circulation as climate changes.
- Soil and vegetation characteristics are represented in the models, and the current hydrological cycle is modeled. However, certain key characteristics (such as vegetation type) do not evolve in the models in response to the simulated climate change.
- Simplified representations of clouds and ice formation and melting are also included.

GCMs have been run using these and other simplified representations of the earth and its current atmospheric composition. Over simulated time, GCMs are able to produce weather and climate patterns that are consistent with observed weather. Seasons occur in the model as in nature, and broad patterns of rainfall are reproduced correctly. From these experiments, it has been possible to develop some reasonable confidence that the models reproduce the climate system, at least for relatively large-scale phenomena.

The models are then used to assess the potential changes in the climate system due to changes in the composition of the atmosphere. In the GCMs, the buildup of greenhouse gases can be represented by altering the estimate of infrared radiation trapped by the atmosphere. This modification influences many elements of the climate system, changing the climate patterns that emerge over many years of simulated weather.

Deficiencies exist in GCMs, however. These deficiencies currently prevent GCM outputs from being usable as climate forecasts

Table 1
Subset of Quantities Used to Describe Weather Simulated by GCMs

Precipitation
Evaporation
Ground wetness
Runoff
Ground temperature
Surface temperature
Surface wind speed
Surface wind direction
Heat flux

at the regional level for atmospheres with compositions that are different from those observed today. Table 2 summarizes some of these deficiencies.

In addition to these modeling deficiencies, researchers may lack the computer resources necessary to exercise their models to their full capacity. Consequently, the sensitivities and full power of the models are not as well understood as desired. GCM experiments require enormous amounts of computer time, with a single experiment often taking months. Without adequate computer resources (including the people to analyze runs), various alternatives for representing climate cannot be tested and much valuable information goes undeveloped.

Reliable climate forecasts will be forthcoming only as these deficiencies are removed. The speed at which this happens will depend, in large part, on how quickly the needed data are collected (very little is known about the oceans, for example) and on the level of support given to scientific researchers. In addition, research must be organized to assure that multidisciplinary teams receive steady funding--funding that starts and stops has been a problem to date.

MODEL ASSUMPTIONS REQUIRED

As discussed earlier, to use GCM results to create hypothetical climate change scenarios, assumptions are required. These assumptions fall into three major areas:

- time distribution of climate change;
- geographic disaggregation of GCM data; and
- the impact of climate change on the interannual variation of weather.

Any scenario will be strongly influenced by the assumptions that form its basis. Each of these assumptions is described below.

Time Distribution of Climate Change

Climate change scenarios must describe how climate variables will change over time, i.e., the transient climate. Scientists are currently uncertain about the transient climate, not only because the time distribution of future greenhouse-gas buildup is uncertain, but also because climate change will probably not progress at the same rate as the buildup of these gases. For example, global warming will lag behind the rate of increase in atmospheric CO_2 because the oceans absorb heat. Also, climate changes within some regions may be faster than the global change as weather patterns alter

Table 2
GCM Deficiencies

	GCMs	Real climate system
Geographic representation	Coarse representation leads to averaging (e.g., mountains and valleys are averaged into a plateau).	The real world's more variable topography would lead to somewhat different climate patterns.
Clouds	Cloud processes are represented for large areas using historical statistical relationships.	Many localized processes may determine cloud responses.
Oceans	Sea surface temperatures change only in response to ocean heat absorption.	Sea surface temperatures will also be altered by changes in ocean circulation.
Hydrological system	Hydrological system is fixed; vegetation and soil do not change in response to climate change.	Hydrological system will evolve; vegetation and soil will change in response to climate change, thereby influencing climate change itself.
Polar regions	Many processes are represented at a highly averaged level.	Many important processes occur in small areas.

their typical behavior. In addition, some areas may change very differently. For example, although an area's climate may become drier in the long term, it may first become wetter before becoming drier. Consequently, transient shifts in precipitation and wind patterns may not necessarily be proportional to the amount of global warming occurring over time.

Figure 1 displays alternative types of potential transient climate change. The figure shows a climate variable increasing over time from point A to point B as a result of increasing concentrations of greenhouse gases. Current GCMs estimate values for A and B (e.g., a temperature increase of 4°C or 7.2°F). However, the models do not assess the path via which the climate variable will move from A to B. A linear path, for example, would follow line 1 in Figure 1. Alternatively, as expected with temperature change, the climate variable may change slowly at first and then at an increasing rate, as shown in line 2. Finally, it has been hypothesized that some climate variables (such as precipitation) may fall outside the interval between A and B before reaching point B (see lines 4 and 5 in Figure 1).

While choosing assumptions for transient scenarios, attention must be paid to spatial consistency. If area 1 is projected to become wetter and then drier as the result of the shifting of a large weather pattern, the implications of this shift must also be examined for other areas. For example, to be consistent, area 2 may be required to become drier and then wetter. Careful examination of the implications of the various spatial and temporal assumptions is required.

For the purposes of the scenario developed in this project, the changes in temperature and precipitation were assumed to be linear (line 1 in Figure 1). All climate variables were assumed to move toward their doubled-CO_2 values at a constant rate over time.[3]

Geographic Disaggregation

For the climate change scenarios to be of value to environmental and economic researchers, they must provide estimates for small

---

[3]In subsequent work we have examined alternative hypotheses regarding the rate of change of climate variables. A one-dimensional transient temperature model is being used to identify the potential rate of temperature change. The rate of change of precipitation is being examined using historical relationships between changes in temperature and precipitation.

Figure 1
Potential Types of Transient Climate Change

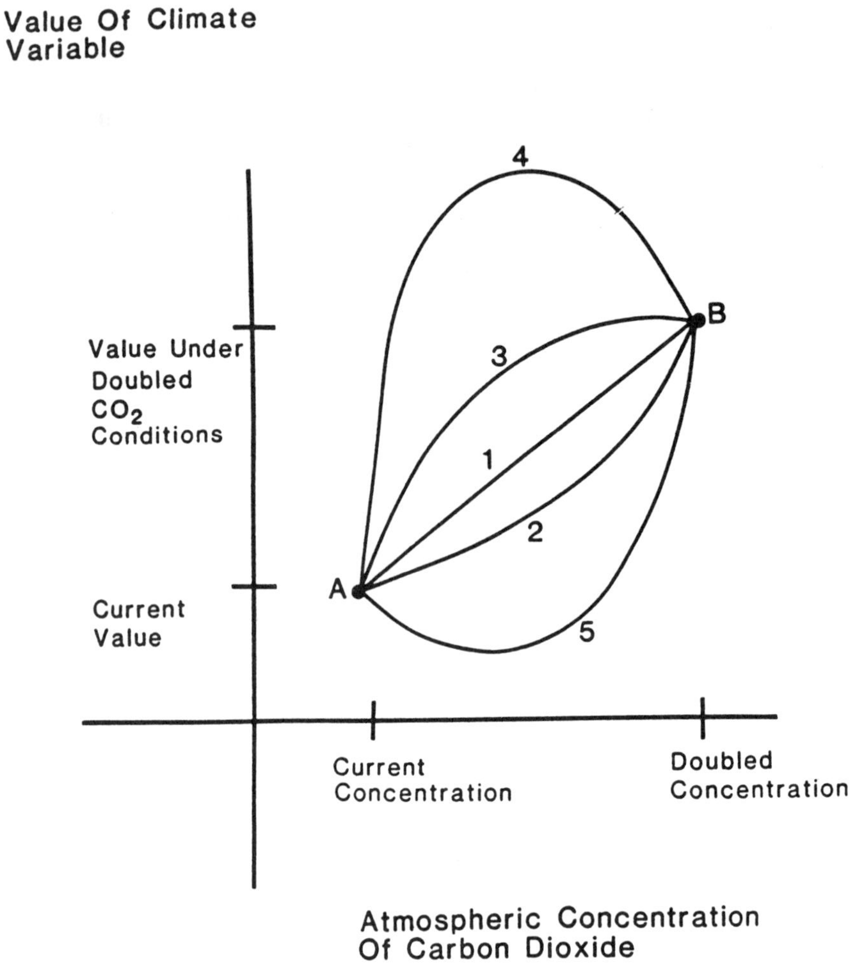

Estimates of points A and B are obtained from GCMs. There are a variety of "paths" via which the climate variables of interest may move from A to B. Shown here are five illustrative paths.

geographic areas. Currently, GCMs have grid boxes (i.e., regions) of approximately 5° latitude by 5° longitude (depending on the model). For some uses, these areas are too large for performing even preliminary impact assessments.

The areas chosen for creating the scenarios in this project are the National Climatic Center (NCC) climatological divisions shown in Figure 2. The method adopted to provide estimates at this divisional level begins with the GCM regional grid cell data (whose boundaries are also shown in Figure 2). Although it is recognized that the GCM regional results are unreliable, they are the best available data for producing the needed estimates at the divisional level. By repeating the analysis using regional results from all the available GCMs, a sense of the disparity among the models can be obtained.

To disaggregate the GCM regional climate change estimates into division-level estimates of climate change, historical relationships between the divisions and the regional areas were developed. The historical relationships, based on 50 years of observation (1931-1980), describe the manner in which the divisional and regional climate data have been correlated.[4] Using the relationship between each division and the region (or regions) it lies in, the expected climate changes at the regional level were modified to produce estimates of climate change at the divisional level.

This approach has several advantages, including:

- objectivity: the method uses all available data in an objective manner;
- local influences: the relationships between division and regions represent local physical influences such as mountains, lakes, and vegetation as they are reflected in weather observations; and
- simplicity: the method is straightforward to implement.

A disadvantage of this approach is that global warming may change the relationships among divisions and regions. Consequently, the historical data used to develop these relationships may not be applicable to account for this possibility.

[4]The historical relationships were developed using divisional temperature and precipitation data from the NCC. Appendix A provides a step-by-step description of how these relationships were developed and used.

Figure 2
Map of NCC Climatological Divisions and Regions from a CGM

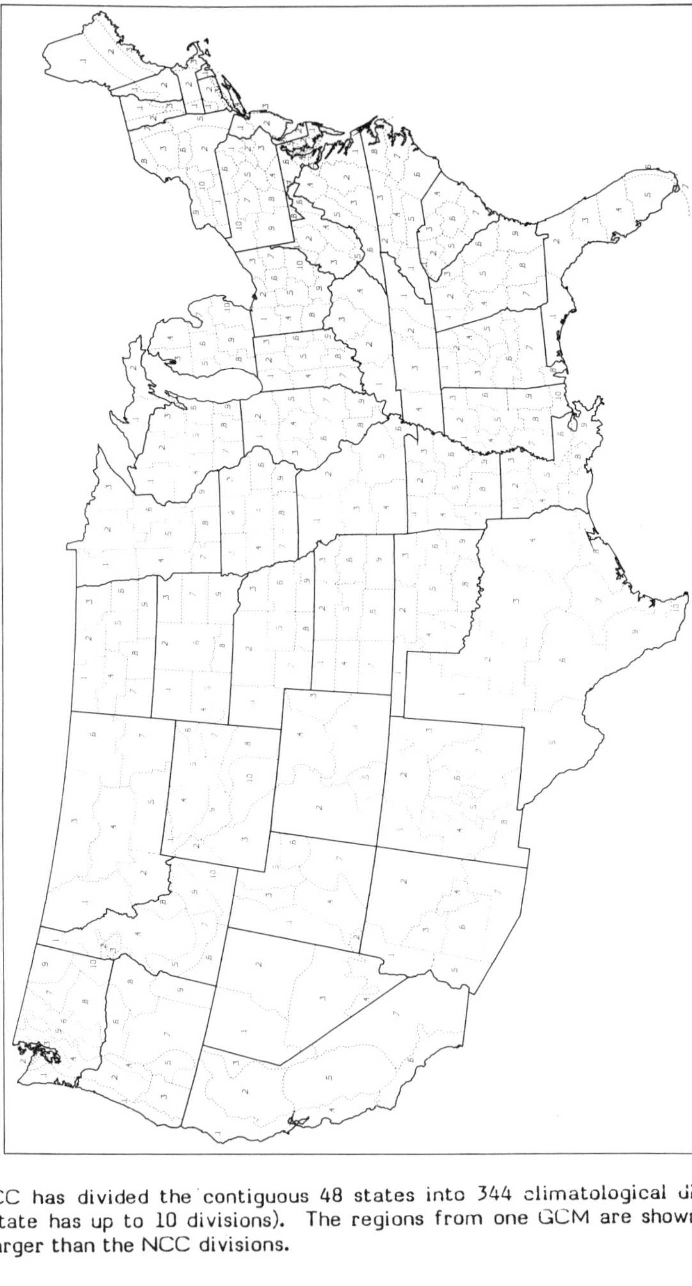

The NCC has divided the contiguous 48 states into 344 climatological divisions (each state has up to 10 divisions). The regions from one GCM are shown to be much larger than the NCC divisions.

Year-to-Year Variability

An important characteristic of climate is its year-to-year variability. Changes in this variability due to global warming may be as important as or more important than the warming itself.

The GCM used to develop the scenarios for this project has not yet provided an estimate of potential changes in interannual variability. Consequently, it was necessary to assume that the variability remains unchanged from historical variability.[5]

SAMPLE RESULTS

Table 3 displays the temperature results from the climate change scenario developed in this project. The results are for region 12,35 (see Figure 2), which includes parts of Colorado, Kansas, Oklahoma, Texas, and New Mexico. In addition, results are shown for divisions 1 and 5 in Colorado and division 2 in New Mexico.

The table shows mean monthly temperatures and changes in mean monthly temperature in degrees centigrade. For example, the GCM estimates the mean January temperature for the region as -7.4°C. Using the historical data for all the divisions in the region, the observed regional mean for January (1931-1980) is -2.2°C. Clearly, the GCM and observed data do not correspond exactly for all months. The third row of the exhibit displays the estimated increase in temperature, 6.1°C for the year. The changes in the monthly values range from 2.7°C (April) to 10.8°C (August).

The second half of Table 2 shows temperature means and changes for three divisions in the region. The means are computed based on historical observations (1931-1980). The changes are computed using the GCM estimates and the historical relationships between the divisions and the region. For example, due to the division-region relationship, the temperature increase in Colorado division 1 is slightly higher than the increase estimated for the entire region (6.2°C versus 6.1°C on an annual basis). The change is magnified even more in Colorado division 5 (6.6°C versus 6.1°C). Finally, in New Mexico division 2, the change is smaller than the change for the entire region (5.6°C versus 6.1°C).

Similar magnification and reduction of changes were identified for all the divisions. The magnification and reduction of the re-

---

[5]Experiments have been, and are currently being, performed to assess potential changes in variability.

Table 3
Temperature Change Scenario for Region 12,35 and Three Divisions

	Temperature (^{o}C)												
	Jan.	Feb.	Mar.	Apr.	May	June	July	Aug.	Sep.	Oct.	Nov.	Dec.	Ann.
GCM estimate of regional mean	-7.4	-7.3	1.8	10.5	13.2	19.5	21.6	21.8	15.5	6.5	-4.1	-7.0	7.1
Observed regional mean	-2.2	0.2	3.6	9.0	14.0	19.2	22.1	21.0	16.9	11.1	3.6	-0.6	9.9
GCM estimate of temperature increase	3.8	10.3	5.1	2.7	4.1	5.5	6.8	10.8	7.8	5.4	6.5	5.1	6.1
Colorado Division 1													
Historical mean	-2.2	-0.2	2.8	8.2	13.3	18.7	21.7	20.6	16.5	10.6	3.2	-0.6	9.4
Change	3.9	10.4	5.2	2.7	4.1	5.5	6.9	10.9	7.8	5.4	6.6	5.2	6.2
Hypothetical mean	1.7	10.2	8.0	10.9	17.4	24.2	28.6	31.5	24.3	16.0	9.8	4.6	15.6
Colorado Division 5													
Historical mean	-8.4	-5.3	-1.0	4.2	9.1	13.7	16.6	15.6	12.0	6.5	-1.2	-6.8	4.6
Change	4.1	11.0	5.5	2.8	4.4	5.8	7.3	11.5	8.3	5.8	7.0	5.5	6.6
Hypothetical mean	-4.3	5.7	4.5	7.0	13.5	19.5	23.9	27.1	20.3	12.3	5.8	-1.3	11.2
New Mexico Division 2													
Historical mean	-3.0	-0.8	2.4	7.2	11.9	16.8	19.4	18.4	14.9	9.4	2.5	-1.5	8.2
Change	3.5	9.4	4.7	2.4	3.7	5.0	6.2	9.8	7.1	4.9	6.0	4.7	5.6
Hypothetical mean	0.5	8.6	7.1	9.6	15.6	21.8	25.6	28.2	22.0	14.3	8.5	3.2	13.8

gional changes at the divisional level can be expressed as ratios. In Colorado division 5, the magnification ratio is 6.6/6.1, or 1.08. In New Mexico division 2, the reduction ratio is 0.92 (5.6/6.1). The full range of reduction and magnification ratios for temperature changes was found to be quite small, approximately 0.8 to 1.2. On the other hand, for precipitation, the magnification was found to be as large as 2.1 and the reduction was as low as 0.4. The larger range was expected for precipitation, which has greater spatial variability than temperature.

NEXT STEPS

Before a complete and useful series of internally consistent hypothetical climate change scenarios can be produced, several additional analyses are required, including:

- *Identification of the most important climate variables.* The current scenario includes only mean monthly temperature and precipitation. Other variables, such as wind direction, ground wetness, and maximum temperature may be desirable.
- *Comparison of the GCM results.* The results of all the GCMs must be compared to identify the full range the scenarios must reflect.
- *Development of a range of assumptions for transient climate and disaggregation.* The scenarios will be sensitive to the assumptions made about transient changes and disaggregation. Internally consistent sets of assumptions must be developed to reflect the likely range of results.
- *Estimation of the interyear variability in the results.* Because interyear variability and the potential changes in the variability are important, a measure of variability must be developed and included as a standard result. One approach would be to simulate weather using probabilistic techniques and to report the frequency with which a certain event occurs (e.g., how often summer precipitation is less than 10 inches). Such estimates would reflect the variability and the potential impacts of changes in variability.
- *Presentation of results in useful formats and media.* The method for presenting the results and making them available to researchers must be improved. Graphic representations are preferred to tabular form, and raw data should be made available on standard computer tapes.

The effort described in this paper clearly constitutes only one step of the many needed to provide hypothetical climate change scenarios that are useful for evaluating the impacts of global warming on various economic and environmental activities.

APPENDIX

This appendix describes the method used to prepare the hypothetical climate change scenario used in the conference Rising CO_2 and Changing Climate.[6] The objective of this work is to produce the following information for each of the climatological divisions defined by the NCC:

- historical monthly temperature and precipitation (mean and standard deviation);
- the expected mean monthly temperature and precipitation under conditions of doubled concentrations of atmospheric CO_2; and
- a description of the manner in which the historical values move toward the doubled-CO_2 values over time.

A variety of analyses may be used to estimate the desired values from the available data. The method used was simple and objective; it required no additional investigation, which was important because of time constraints. The analytic approach is composed of four parts; each is described below.

I. Data

The following data were used:

- Historical NCC divisional climate data: 50 years (1931-1980) of observed monthly precipitation and temperature for each of the 344 NCC climate divisions.
- GCM regional climate data: an estimate of monthly precipitation and temperature under current climate conditions (control data) and under doubled-CO_2 conditions (doubled-CO_2 data) for a given set of regions which cover the continental United States.
- Mapping data: prepared by using maps to identify the NCC divisions in each GCM region.
- Historical region-equivalent climate data: computed in part II of the analysis and stored (see below). Using the historical NCC divisional data and the mapping data, historical data

---

[6]Sponsored by the National Forest Products Association, Society of American Foresters, and The Conservation Foundation, June 25-27, 1984, Boulder, Colo.

for the region-equivalents were calculated by areally weighting the divisional observations within the region-equivalents.

II. Analysis of Regions and Region-Equivalents

The purposes of this part of the anlysis were to:

- compare the control and doubled-CO_2 GCM data and calculate the GCM-estimated climate changes;
- prepare and summarize the historical region-equivalent data; and
- compare the control GCM data and the historical region-equivalent data and calculate region-equivalent climate change.

These computations were performed using Fortran programs developed by ICF Inc. The analysis proceeded in the following four steps:

- *Step 1*: Compare the control and doubled-CO_2 GCM data to estimate GCM "deltas" (Δs). Deltas are defined as the changes in the monthly means that are attributable to the doubling of the concentration of CO_2. An output table was produced showing the control data, the doubled-CO_2 data, and the estimated deltas.
- *Step 2*: For each region-equivalent, monthly temperature and precipitation values were computed using the mapping file and the NCC divisional data for the full 50-year period. These results were retained as the historical region-equivalent climate data.
- *Step 3*: For each region-equivalent, the mean monthly precipitation and temperature were estimated from the historical region-equivalent climate data.
- *Step 4*: Deltas were calculated for the region-equivalent areas using the deltas computed from the GCM data. To take into account the fact that the GCM control data do not approximate the historical data in all cases for precipitation, the changes in precipitation were made on a *percentage* change basis. The region-equivalent deltas were computed as:
  - -- Monthly mean temperature delta = delta computed from GCM data
  - -- Annual mean precipitation delta = percent change in annual precipitation from the GCM data, times the historical annual average precipitation in the region-equivalent area

-- Monthly mean precipitation delta = annual mean precipitation delta apportioned across the months according to the apportionment in the GCM data

III. Analysis of Specific NCC Divisions

The purpose of this part of the analysis is to transform the region-equivalent deltas (estimated in part II) into division-specific deltas. The analysis is performed for each division.

This portion of the analysis was implemented on the ICF computer separately from the other evaluations. It required the use of SAS (Statistical Analysis System) to perform and evaluate regressions of the historical NCC divisional data and the historical region-equivalent data. The approach has several obvious weaknesses, the chief among them being serial correlation among the independent variables (i.e., the historical data) used in the regression. Various efforts could be undertaken to improve the method; however, because of time constraints, the following four steps were performed.

Step 1. Using SAS, regress 50 years (i.e., 600 months) of data as follows:

$$[D]^T = A_0^T + A_1^T[RE_1]^T + A_2[RE_2]^T + \ldots$$

where D^T is the division temperature data and $RE^T{}_1$ is the region-equivalent temperature data. The values of the coefficients, A_1^T are for temperature and are stored. (Analogous $A^P{}_1$ values for precipitation were also calculated.) The A_1 values can be considered as "weights" or "magnification/reduction factors" for the region-equivalent deltas when computing the division-specific deltas.

Step 2. Compute division deltas using:

$$\text{division } \Delta = A_1 \ (RE_1 \ \Delta) + A_2(RE_2 \ \Delta) + \ldots$$

where $RE_1 \ \Delta$ is the region-equivalent monthly deltas computed in part II.

Step 3. Compute the historical mean monthly temperature and precipitation for the division of interest.

Step 4. Create an output table showing the results of steps 2 and 3. The output includes monthly historical values, monthly deltas, and monthly historical values plus monthly deltas (i.e., monthly doubled-CO_2 values). The table can be interpreted as the current climate in the division and the change expected as the result of a doubling of CO_2.

IV. Interpolate over Time

The purpose of this part of the analysis was to interpolate between the historical and doubled-CO_2 estimates made in step 4 of part III for a given division. This portion is a third computer program. The routine interpolates over time according to user-specified inputs. Linear interpolation was used for this analysis, and it was assumed that the change in climate would require 100 years to occur.

The magnitudes of changes in annual mean temperature and precipitation produced using this method are shown in Figures 3 and 4.

Figure 3

Changes in Mean Annual Temperature
due to a Doubling of Atmospheric Carbon Dioxide (°C)

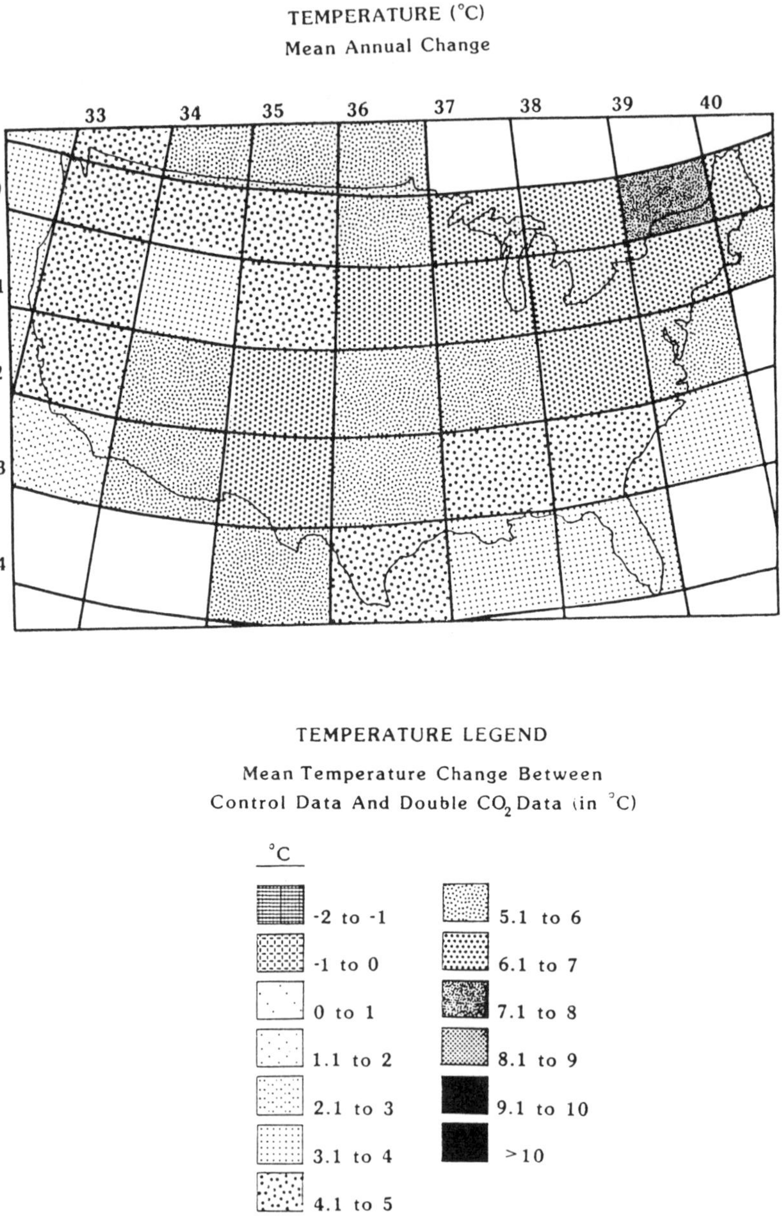

Figure 4

Changes in Mean Annual Precipitation due to a Doubling of Atmospheric Carbon Dioxide (%)

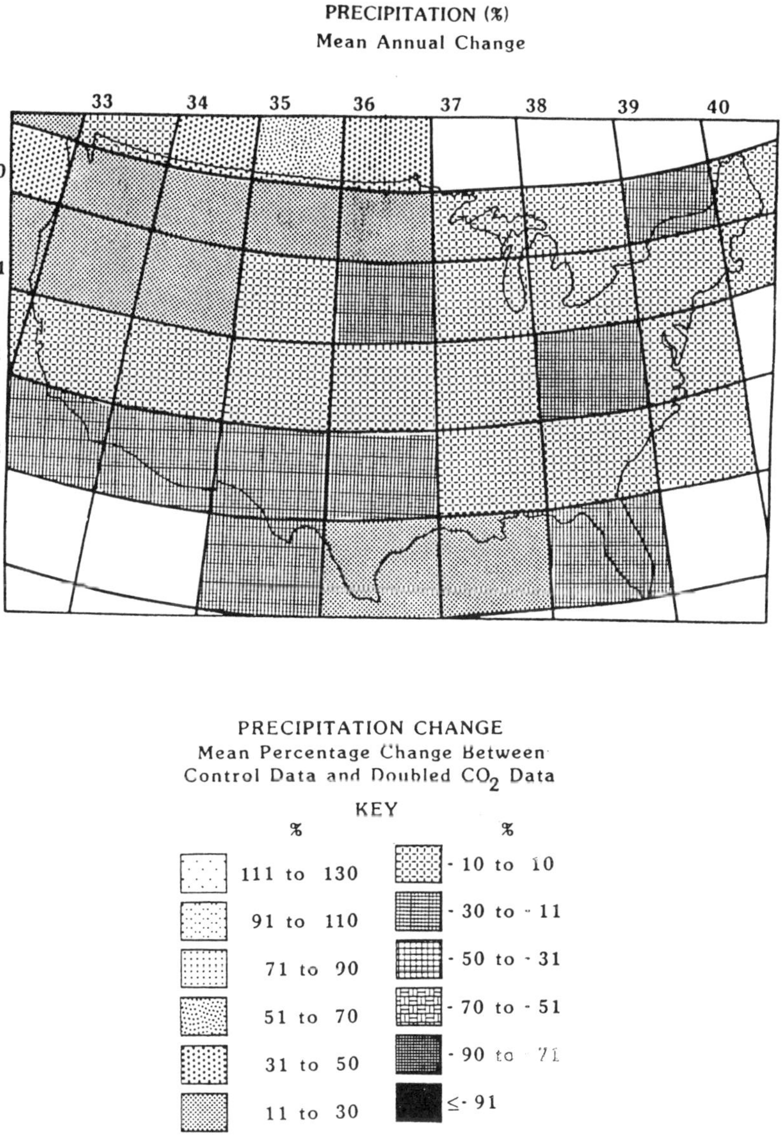

RISING CARBON DIOXIDE, CLIMATE CHANGE, AND FOREST MANAGEMENT: AN OVERVIEW

by Robin Sandenburgh, Carol Taylor, and John S. Hoffman[1]

ABSTRACT

Rising carbon dioxide (CO_2) will alter the biochemical processes of trees and plants, changing growth rates and productivity. Moreover, environmental change resulting from increased atmospheric CO_2 could alter the potential productivity of the world's forest regions. Changes are likely to affect timber-growing regions of the United States in different ways, particularly intensifying extremes in temperature, rainfall, and wind. Changes in the environment over the life of a stand could exceed those of the past century; trees being planted today could experience the most productive parts of their lives in conditions quite different from today's.

An accumulation of atmospheric CO_2 could directly alter biochemical cycles in ways which influence the distribution of species and forest productivity. These include change in the rate of photosynthesis, change in water-use efficiency, and increased nitrogen fixation. In addition to biochemical changes, there is likely to be increased risk of loss to fire and pests.

To maintain or enhance forest productivity, adjustments will be required in forest technology, resource allocation, planning, and decision making.

If scientific theories about the effects of rising levels of carbon dioxide (CO_2) prove correct, the planet's average temperature will increase significantly over the next century, changing regional climatic conditions (National Academy of Sciences, 1979). Rising atmospheric CO_2 will alter biochemical processes of trees and plants, changing both growth rates and productivity (Wittwer,

---

[1]This material is adapted from the report *Strategic Options for Enhancing Forest Industry Productivity in the Face of Rising Atmospheric Carbon Dioxide*, prepared by ICF Incorporated for the U.S. Environmental Protection Agency.

1978). Together these changes will disrupt ecosystems and change growing conditions of forested regions throughout the world, with decade-to-decade changes occurring faster than at any time in the history of managed forests and with cumulative changes over several decades vastly exceeding those of the past. These environmental changes could alter the potential productivity of the world's forest regions. Actual forest productivity will depend, however, not just on these environmental changes but on the responses of forest managers.

If the National Academy of Sciences' panel on CO_2 is correct, cumulative changes in the environment over the life of various tree species will exceed those of the past century. Trees being planted today will experience the most productive part of their life cycles in conditions quite different from today's.

Changes in local climate associated with global warming could include shifts in:

- average precipitation, evaporation, and soil moisture;
- the frequency of extremes such as the likelihood of droughts or floods;
- the total variability of climatic conditions faced over the life of a stand;
- seasonal characteristics, such as first or last frost or summer dryness; and
- frequency of damaging storms (Manabe et al., 1981).

The changes will not be the same everywhere: the climate in some regions could become warmer and wetter and in others, warmer and drier. The temperature will increase more in higher latitudes, less in equatorial areas.

In addition to climate effects, rising atmospheric CO_2 amounts may also increase plant growth by affecting photosynthetic rate and plant physiology. Higher levels of CO_2 may lower plant water requirements and allow soil microbes to "fix" greater amounts of nitrogen (Hardy and Havelka, 1975). Specific effects will vary among tree species as well as among varieties and populations within a species. These effects will also depend, in part, on changes in climate. At present, the extent of biochemical and physiological changes in different species and under different environmental, climatic, and management conditions is far from clear. The evidence, however, indicates that productivity will be higher than without the CO_2 fertilizer effect and that forest companies can enhance possible increases by selecting the most responsive species and by appropriately managing other resource inputs. Whether or not the combined effects of climate change and CO_2 fertilization will be positive or negative is unclear; it is certain, however, that prior planning can reduce losses or enhance increases as the environment changes.

Planning for Climatic and Biochemical Change

The combined effects of climatic and biochemical change could influence the potential productivity of key silvicultural regions. The productivity of a given forest region will be determined by how particular species grow over an entire life cycle. This, in turn, is a function of natural growing conditions over time, technologies and resources available, and decisions to apply these management options over the long haul. If CO_2 induces changes in natural growing conditions, adjustments will be needed in forest technology, resource allocation, planning, and decision making to maintain or increase productivity.

Many factors combine and interact to determine growth, so projection of ultimate effects is difficult. Because CO_2 and other greenhouse gases have not accumulated nearly as much in this century as they will in the next, direct experience is insufficient to evaluate what will happen. Using evidence from past changes in climatic conditions nevertheless offers one method of generating hypotheses about how CO_2-induced change could influence growing conditions and overall productivity. A second method is to look at how productivity varies with different growing conditions found in existing climate regions. Neither analogue method, however, accounts for the dynamic dysequilibrium that CO_2 could induce. No region existing today has experienced changes in year-to-year and decade-to-decade conditions equal in magnitude to those that could occur under rising CO_2 nor have any regions simultaneously experienced equivalent changes in biochemical and physiological processes.

Regional Climatic Change

Because researchers studying CO_2 buildup agree that regional climates will change more than global averages, the concentration of important forests in certain regions is important. In the United States, three regions, the South, North, and Pacific, account for 95% of the nation's annual timber production (U.S. Department of Commerce, 1976). Worldwide timber holdings also are concentrated regionally. It is unlikely, however, that all regions would either improve or worsen; it is equally unlikely that each, or any, would be influenced in the same way.

Distinct natural distribution patterns for most species of trees are the clearest evidence that tree growth depends on climate. Cypress trees do not flourish in Minnesota, Douglas-fir is not apt to be highly productive in Florida, and no trees grow in the Sahara Desert or on Antarctica.

Silvicultural management decisions recognize the importance of local weather and climate parameters to plant growth and output. Managers choose tree species and varieties for the climate of an area, including expected climate variations over a stand's lifetime. Yet, if climatic norms and expected variances are so important to tree growth, what will happen as these norms and variances shift continually and in one, warm, direction? Layser (1980) reviewed historical studies on the influence of regional climate on forest productivity and concluded that:

> ... it would appear that climate change may significantly affect projected yields and rotation length for forest tree species, and might have a bearing on the concept of sustained high-yield forestry....Since trees keep their genetic make-up throughout a long life...forest managers may need to take climatic trends into account in silviculture, and view site as a dynamic, not static, concept.

The principles of silviculture suggest that changes in weather extremes could be important to forest growth. CO_2-induced climate change could alter current extreme weather conditions either way, producing greater or fewer stressful periods. Of the fundamental climate factors--wind, temperature, and rainfall--rainfall is the single strongest influence on growth.

Temperature Extremes

Although higher levels of atmospheric CO_2 are expected to have a warming effect overall, average temperatures will vary regionally, and extreme temperature conditions could increase or decrease in some regions. For example, if an area experiences more "cold snaps," higher tree mortality rates may result. The surviving trees may suffer loss of vigor and show subsequent decreases in budding. As a result, the natural regenerative capacity of the forest would be reduced. Conversely, fewer cold snaps in northern areas would reduce such problems there.

In some areas, trees exposed to more heat waves could suffer short-term water stress, diminished vigor, and enhanced vulnerability to external attacks by insects and disease. Conversely, an area now subject to heat waves might have fewer because of changing circulation patterns, decreasing this stress.

Precipitation Extremes

Most forest species are resilient enough to withstand occasional drenching rains, extreme dry periods, and even short-term inundation without experiencing significant losses. Numerous precipitation extremes--in either direction--will reduce productivity of the species being planted now, however. More frequent wet conditions over time could increase insect and pest populations and, by increasing soil moisture, enhance potential root damage. Too frequent dry conditions in a given period, on the other hand, could lead to water stress. Again, the converse holds: less-productive regions may be favored by changes in drought or flood occurrences.

Wind Extremes

Particularly in the critical first stage of forest establishment, extreme winds may be crucial. Seedlings grown on level, windswept, open-field sites have shown less vigorous growth, reduced leaf area, and greater foliage damage than those grown under similar soil conditions but in forest openings or protected fields (Spurr and Barnes, 1973).

From a management viewpoint, wind is a force because it is potentially destructive. Few forests can stand the devastating wind velocities of a hurricane or tornado, but some species are better suited than others. An area never hit by hurricanes likely is planted with a species ill-suited to gale winds. Beyond the destruction from major windstorms, however, is breakage attributed to lesser, "normal" winds. Fewer windstorms of course would improve the potential productivity of an area and open it to new species.

Forest Fire Risk

In regions where average temperature increases and average rainfall drops, the risk of fire goes up. In the last decade, annual fire losses in National Forests and adjacent areas alone amounted to more than $100 million (U.S. Forest Service, 1980). If summers become drier, the work of Manabe et al. suggests that fire becomes a powerful threat (Manabe et al., 1981).

Even though advances in forest fire management since the early part of the century are impressive, small changes in climate can overshadow any progress. In the last decade, the incidence of fires went up (Nelson, 1979). Data on the relationship of climate to fires in Canada and the United States show the weight of climate (U.S. Department of Agriculture, 1981). The 4.8 million--hectare area that burned in 1980 was 28% greater than the area burned in 1961, the previous record year, prompting the Canadian government to report that:

> Preliminary data support the idea that an upward trend in burned area is linked both to the increased use of the land by people, and to worsening climatic trends during the past two to three decades (U.S. Department of Agriculture, 1981, emphasis added).

Losses to Pests

Millions of dollars are spent annually by forest managers to control insects, weeds, bacteria, fungi, and predators. Despite improved pest control, current losses average about 25% of total output (Pimental *in* U.S. Department of Energy, 1980).

Fungi and insects are especially dependent on temperature. The well-being of pathogens, fungi, bacteria, and weeds is tied to water availability. Moreover, warmer temperatures could change the range and geographical region of insects during winter and, hence, the areas in which niches can be established for the next growing season. Weather patterns and pest outbreaks correlate. For example, the 1912 and 1949 spruce budworm outbreaks were statistically associated with dry, warm fall weather patterns (Haynes, 1984).

Whether pests increase or decrease, they will nevertheless be affected by climatic alternations. As one author put it:

> ... insects and plants are evolving, competing, interdependent systems. Because they have evolved together for eons, the balance between them may be very fine and easily, unwittingly, unrecognizably, and irrevocably altered by our actions (Stark, 1977).

CHANGES IN PLANT PHYSIOLOGY

Trees and other plants establish niches in an ecosystem and grow and compete on the basis of biochemical and physiological processes that determine their competitiveness with other organisms. Accumulated CO_2 can directly alter biochemical cycles and physiological processes. Both theoretical and empirical evidence suggests that rising CO_2 levels can increase plant photosynthesis rates differentially and alter other biochemical cycles in ways that may influence the productivity of trees and plants, as well as their competitive distribution in natural ecosystems.

Photosynthesis

Carbon dioxide is essential to photosynthesis, the process by which plants convert water, sunlight, and CO_2 into sugars. Scientists who study photosynthesis believe that plant growth depends, in part, on CO_2. Evidence suggests that current ambient levels fail to provide sufficient CO_2 to permit maximum growth (Baker et al., 1983). If this is so, a rise in the ambient CO_2 available to plants could increase the rate of photosynthesis and plant growth, as long as other factors--water, nutrients, solar radiation, or internal plant processes--do not limit growth (Lemon, 1977). In other words, the level of CO_2 now available in the atmosphere may be stunting growth. More CO_2 may help trees grow faster (see Kramer and Sionit, this volume, for a full discussion).

Species are likely to differ in their responses to greater CO_2 exposure, as do different parts of populations within the same species. Despite observed differences, however, few of the physiological or biochemical factors that account for such significant differences really are understood. A review of many studies suggests several factors critical to determining plant responses to higher CO_2:

- the dependence of the plant on nitrogen for growth;
- the indeterminateness or determinateness of a species;
- the rate at which a plant grows during critical phases in its development; and
- the plant's ability to survive the stresses of each life-cycle stage as a result of numerous environmental and climatic factors (Tinus, 1974; Leidig, 1976; Allen, 1979; Kramer, 1981).

Biochemical and Physiological Processes

Scientists have studied the influences of higher ambient CO_2 levels on plant growth and productivity. Higher CO_2 concentration appears to improve water-use efficiency by decreasing stomatal aperture. Other work showed that an increase in CO_2 concentration of 250 parts per million above current levels improved both water-use efficiency and yield in water-stressed wheat compared to similar plants grown under normal CO_2 concentrations (Gifford, 1979).

Similarly enhanced nitrogen fixation was demonstrated in experiments growing soybeans under higher CO_2 levels (Hardy and Havelka, 1975): a threefold increase in ambient concentration resulted in a fivefold increase in nitrogen fixation.

Other effects demonstrated in experimental studies include greater branching, elongation of stems, increase in root/shoot ratios, larger leaf area and, in some cases, thicker leaves, and a hastening of bud initiation and flowering (Hesketh and Hellmers, 1973). These effects have been observed in only a few selected species; thus it is incorrect to draw general conclusions from these results.

The key issue for commercial silviculture is the extent to which increased photosynthesis and changes in other biochemical and physiological processes translate into increased yield over a tree's life. Evidence from plants grown in greenhouses and under experimental conditions, in which other factors (water, soil, and nutrients) are not limiting, does suggest that higher CO_2 levels could increase plant yield significantly. Conclusions of how much, however, are far from definitive--especially considering particular species grown under actual field conditions.

REFERENCES

Allen, L. H. 1979. Potentials for carbon dioxide enrichment. *In* B. J. Barfield and J. F. Gerber (ed.), *Modification of the Aerial Environment of Crops*. American Society of Agricultural Engineers, St. Joseph, Mich.

Baker, D. N., et al. 1983. The impact of increased CO_2 on photosynthesis and agricultural productivity. Mississippi State University, Mississippi State.

Gifford, R. M. 1979. Growth and yield of CO_2-enriched wheat under water limited conditions. *Aust. J. Plant Physiol.* 6, 367-378.

Hardy, R. W. F., and U. D. Havelka. 1975. Nitrogen fixation research: a key to world food. *Science* 188, 633-643.

Haynes, D. L. 1984. Effects on agricultural plant pests. Department of Entomology, Michigan State University, East Lansing.

Hesketh, J. D., and H. Hellmers. 1973. Floral initiation in four plant species growing in CO_2-enriched air. *Environ. Control Biol.* 11, 51-53.

Kramer, P. J. 1981. Carbon dioxide concentration, photosynthesis, and dry matter production. *Bioscience* 31, 29-33.

Layser, E. F. 1980. Forestry and climatic change. *J. For.* November, 678-682.

Ledig, F. T. 1976. Physiological genetics, photosynthesis and growth models. *In* M. G. R. Cannell and F. T. Last (ed.), *Tree Physiology and Yield Improvement.* Academic Press, Inc., New York. p. 212-54.

Lemon, E. 1977. The land's response to more carbon dioxide. *In* N. R. Anderson and A. Malahoff (ed.), *The Fate of Fossil Fuel CO_2 in the Oceans.* Plenum Publishing Co., New York.

Manabe, S., R. T. Wetherald, and R. J. Stouffer. 1981. *Summer Dryness Due to an Increase of Atmospheric CO_2 Concentration.* National Oceanic and Atmospheric Administration, Fluid Dynamics Laboratory, Washington, D.C.

National Academy of Sciences National Research Council. 1979. *Carbon Dioxide and Climate: a Scientific Assessment.* Report by the Climate Research Board. National Academy of Sciences, Washington, D.C.

Nelson, T. C. 1979. Fire management policy in the national forests--a new era. *J. For.* November, 723-725.

Spurr, S. H., and B. V. Barnes. 1973. *Forestry Ecology*, 2nd ed. The Ronald Press Co., New York.

Stark, R. W. 1977. Integrated pest management in forest practice. *J. For.* May, 251-254.

Tinus, R. W. 1974. Impact of the CO_2 requirement on plant water use. *Agric. Meteorol.* 14, 99-112.

U.S. Department of Agriculture. 1981. More on Canada's 1980 forest fire season. *Forest Fire News*, July.

U.S. Department of Commerce, Bureau of the Census. 1976. *Census of Manufacturers, 1972. Volume II. Industry Statistics. Part I. SIC Major Groups 10-26.* U.S. Government Printing Office, Washington, D.C.

U.S. Department of Energy. 1980. *Workshop on Environmental and Societal Consequences of a Possible CO_2-Induced Climate Change.* Report CONF-7904143 UC-11, October, 1980. U.S. Department of Energy, Washington, D.C.

U.S. Forest Service. 1980. *Annual Fire Report: 1979.* U.S. Government Printing Office, Washington, D.C.

Wittwer, S. H. 1978. Carbon dioxide fertilization of crop plants. *In* U.S. Gupta (ed.), *Crop Physiology.* IBH Publishing Co., Oxford.

EFFECTS OF CARBON DIOXIDE-INDUCED CLIMATE CHANGES ON THE NATURAL RANGES OF SIX MAJOR COMMERCIAL TREE SPECIES IN THE WESTERN UNITED STATES

by Jerry W. Leverenz and Deborah J. Lev

ABSTRACT

Based on knowledge of plants' responses to increased temperature and decreased site water balance, combined with a climatic scenario assuming a doubling of atmospheric CO_2, changes in the natural range of six major forest species were detected. Ponderosa pine increased in area and importance in California and in Oregon's Cascade Mountains. It decreased in range on the east slope of the Rocky Mountains from Canada to Mexico. Douglas-fir either maintained or expanded its importance throughout most of its commercially significant range. However, Douglas-fir decreased in importance on southern and coastal sites and on the east slope of the Rocky Mountains. Western hemlock's range decreased in northern Idaho and east of the Willamette Valley in Oregon. Western hemlock increased in importance along Oregon's coast. Western larch generally increased in importance but decreased in area in Oregon and Washington. The net effect on lodgepole pine was insubstantial. Engelmann spruce declined in acreage over its whole range. These projections were very sensitive to assumed responses to the balance between precipitation and evaporation on a site and the assumed chilling requirements of different species.

By the time one or two more crops of conifer trees can be grown, a minor component of the atmosphere will have doubled in concentration. This minor component, carbon dioxide (CO_2), makes up less than one-thousandth of the atmosphere. CO_2 is difficult to measure accurately without sophisticated laboratory equipment (Gammon, this volume), yet changes in atmospheric carbon dioxide have the potential to affect both directly and indirectly the natural distribution of commercial forest species. Several comments made during the conference apparently indicate that the response of vegetation to a change in climate will significantly alter the distribution of precipitation. Thus as global climate models improve, fundamental knowledge of how trees respond to their environment may become as important as knowledge of atmospheric physics. While this paper considers only the effects of climate on tree distribution, tree distribution may also be affecting climate.

The projections in this paper are based on a climatic scenario that assumes a doubling of atmospheric CO_2 and its attendant climate changes (Gibbs and Hoffman, this volume). These are just projections, not predictions. The reader should take care not to fall into the trap of thinking that the climate changes or plant responses described here are going to occur exactly as presented. Although the scenario is as realistic as possible, there is considerable uncertainty about the degree of global warming and regional climate changes, as well as deficiencies in present understanding of how trees function. Because of these uncertainties the distribution outcomes are projections, not predictions. To produce these projections, several tenuous assumptions were made. One is that the climatic scenario is correct. However, uncertainty exists in the range and types of climatic change; therefore, this paper indicates rather than informs. A second uncertainty, of equal importance, exists in the assumptions made about plant responses to a doubled-CO_2 climate. Therefore, these projected changes should also be taken only as an indication of the variety of responses the species will make rather than a statement of any particular species' fate. Further, coupled with explicit assumptions is the hidden one that anything yet uncovered is unimportant. In the complexity of competition in the natural world, it is easy to ignore some phenomenon that has a major effect on the outcome of the distribution of species.

This study projects distribution outcomes for six western United States tree species--ponderosa pine, Douglas-fir, western hemlock, western larch, lodgepole pine, and Engelmann spruce--derived from available information on physiological changes in plants induced by accumulating (doubled) CO_2 combined with rising temperature and water stress.

PHYSIOLOGICAL RESPONSES TO RISING ATMOSPHERIC CO_2

The interaction between CO_2 and other environmental factors on plant function is paramount. For example, it would be wrong to base projections solely on the response of plants to temperature changes or solely on their responses to changes in atmospheric CO_2.

Temperature

Between -5° and 45°C (23°F and 113°F), CO_2 and temperature are mutually compensating factors for photosynthesis. Increased CO_2 compensates for temperatures above and below the optimum, and temperatures near optimum compensate for low atmospheric CO_2. CO_2 compensates more for supraoptimal temperatures than for suboptimal temperatures (Figure 1). As a consequence, the temperature optimum for photosynthesis shifts upward by 5° to 10°C (9° to 18°F) with the doubling of atmospheric CO_2 to a concentration of 600 parts per million (ppm). Along with a rise in optimum temperature, the absolute rate of net photosynthesis increases at any given temperature (Figure 1). To summarize, the doubling of atmospheric CO_2 more than compensates for temperature increases of up to 10°C (18°F) with respect to photosynthesis.

However, rising temperature will also speed up respiration of nonphotosynthetic plant parts. The rise in respiration with increasing temperature (T) is exponential and can be expressed:

$$R(T) = R(0)\exp(BT)$$

where R(T) is respiration at temperature T; R(0) is respiration at 0°C (32°F); and B is relatively conservative at about 0.07 to 0.08 (Jarvis and Leverenz, 1983). R(0) varies a great deal between species and within individuals during acclimation to season changes in temperature (Berry and Raison, 1981). By using a value of 0.0075 for B, every 5°C (9°F) increase in temperature raises respiration by a factor of 1.45. Thus, despite a large increase in photosynthetic efficiency with doubled atmospheric CO_2, its effect on the balance between carbon uptake and loss for the whole plant will diminish because of the added response of respiration to temperature. Therefore, for the balance between carbon uptake and loss, the increased CO_2 is assumed to compensate for temperature increases of up to about 5°C (9°F).

No one has determined the extent to which the six species of this study acclimatize to higher growth temperatures or higher atmospheric CO_2 concentrations by changing R(0). Some species can acclimate with large changes in R(0) whereas others cannot (Berry and Raison, 1981). Increased atmospheric CO_2 has an impressive ability to augment survival of plants grown at high and low killing temperature (Kramer and Sionit, this volume). Doubled atmospheric CO_2 is not expected to compensate for temperatures above 54°C (130°F), at which point proteins denature.

Chilling Requirements

Early, rapid shoot growth and germination are critical to the survival of trees growing in regions of summer drought or at high altitudes with short growing seasons (Cleary et al., 1978). Rapid

Figure 1

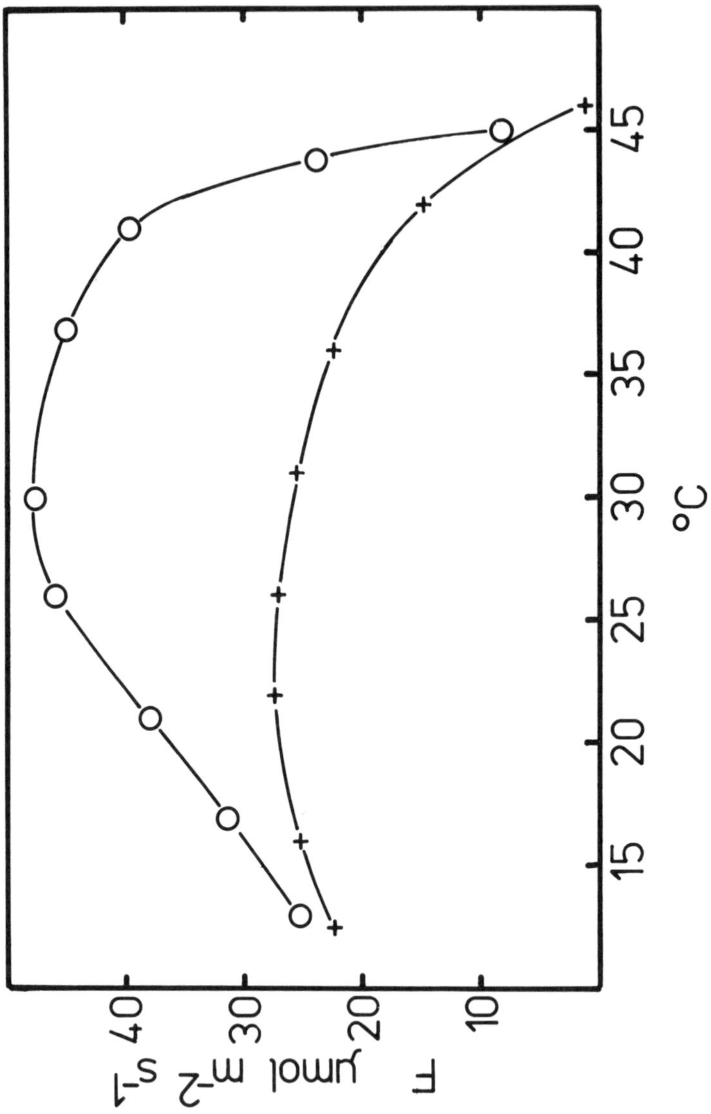

Response of net photosynthesis (F) to leaf temperature at 1000 ppm (O) and 340 ppm (+) CO_2 in air. Redrawn from Berry and Raison (1981).

shoot elongation or germination synchronized to season change depends on trees or seeds receiving sufficient winter chilling (Campbell, 1978; Lavender, 1981). Lavender (1984) noted the potential problem of insufficient chilling in a doubled-CO_2 climate. Like carbon balance, shoot growth and germination are affected by mutually compensating factors. For example, warmer spring temperature can compensate for warm winter temperatures, and increased winter chilling can compensate for cool spring temperatures (Lavender, 1981).

To predict the effect of changing climate on the timing and extent of shoot growth, a comprehensive model is required for each species (L. H. Funchigami, personal communication, 1984). At present no one has studied the direct effect of increased atmospheric CO_2 on chilling requirements. However, at least one reported failure of Douglas-fir resulted from unmet chilling requirements (Copes, 1983). Because Copes noted a failure when mean monthly temperatures did not drop below 9.3°C (49°F), it is assumed that Douglas-fir will be limited to upslope sites in climate divisions where mean temperatures remain above 9°C (48°F).

CO_2 concentrations at the soil's surface are already so high that a doubling of atmospheric CO_2 should not significantly affect seed germination. Chilling requirements for germination of genotypes growing in areas of warmer winters are greater than those for mountain or inland genotypes (Cleary et al., 1978; Lavender, 1981). For coastal zones, where stratification requirements are longest, warmer winter temperatures are more likely to be insufficient to insure proper timing of seed germination to avoid summer drought. However, for lack of information and because some seed always germinates even with stratification, the same 9°C (48°F) cutoff is assumed for germination.

Light

It is often assumed that the rate of photosynthesis is only affected by the partial pressure of CO_2 at saturating light. However, at all light-flux densities, CO_2 and light are mutually compensating factors. On the initial slopes of the light-response curve of photosynthesis, a doubling of atmospheric CO_2 will increase photosynthesis by 25% at 30°C (86°F) (Ehleringer and Bjorkman, 1977). This compensation of CO_2 at low light may have a relátively large effect on net photosynthesis (Kramer and Sionit, this volume). The current assumption here is that cloud cover will change insignificantly if atmospheric CO_2 is doubled. No change in shade tolerance relative to other species is assumed, because any compensation of doubled atmospheric CO_2 on the relative tolerance to shade in the six species in this report is as yet undiscovered.

Water Stress

Photosynthetic water-use efficiency should at least double with a doubling of atmospheric CO_2 content (Baker and Enoch in Lemon, 1983) as a result of water use decreasing by 36% and plant growth increasing by 33% (Baker and Enoch in Lemon, 1983). In some cases a two- to threefold increase in growth occurs when drought strongly suppresses productivity (Baker and Enoch in Lemon, 1983). Accumulated atmospheric CO_2 decreases osmotic potential, which should increase both turgor and the ability of plants to endure water stress and subzero temperature (Pearcy and Bjorkman in Lemon, 1983). Doubled atmospheric CO_2 in general increases root growth in plants (Tinus, 1972; Lemon, 1983). However, in at least one study of Douglas-fir, increased root growth was not observed with elevated CO_2 (Purohit and Tregunna, 1976). Insufficient growth of seedling roots in deep, wet soil is a major reason for drought-death in young conifers (Barrett, 1980).

The water balance on any given site may be roughly modeled as:

$$S = P - E$$

where E is evaporation and P is precipitation. By virtue of the roughly 36% decrease in transpiration, heightened ability to endure water stress, and possibly better root growth, doubled atmospheric CO_2 is assumed to more than compensate for a decreased site water balance when S decreases by less than 25%. Further, doubled CO_2 is assumed to completely compensate for a decrease of S in the range of 25 to 50% and to incompletely compensate for a decrease in S of greater than 50%.

No experiments have tested how the six species considered here will respond to long-term increases in atmospheric CO_2 while under environmental stress. In short-term experiments (hours), stomata of conifers did not respond to changes in atmospheric CO_2 levels (Jarvis and Morison, 1981). This behavior differs strikingly from many broad-leaved trees (angiosperms). If stomatal closure does not occur under chronically high CO_2 (Kramer and Sionit, this volume), projections would be for a reduction in conifer forests almost everywhere. On the other hand, if stomatal control occurs, conifers may have better water-use efficiency than do broad-leaved species, because, unlike broad leaves, conifer needles do not heat up substantially when transpiration slows. This rise in temperature tends to drive broadleaf trees' transpiration back up to where it was at lower atmospheric CO_2 concentrations (Morison and Gifford, 1984).

In general, a doubled atmospheric CO_2 level seems to compensate for every plant stress over fairly broad limits (Lemon, 1983). The primary physiological effects of a doubled CO_2 climate have been covered; however, secondary and tertiary effects have not because no data exist from which to draw a discussion. For example, rapid leader elongation may predispose trees to wind damage. Additional warmth in the wet autumns may prevent hardening off, leaving trees more vulnerable to frost damage. And a potential for reduced nutrient cycling in the long run exists with doubled atmospheric CO_2 (Strain and Bazzaz in Lemon, 1983).

To project the relative responses of the six different species, it is assumed that plants exhibiting the highest rates of growth under favorable conditions will have the largest response to rising levels of atmospheric CO_2 at better sites.

Competition with shrubs or grass may substantially change and severely restrict natural regeneration in some areas. Plants with free growth may hold a big competitive advantage over the conifers with fixed growth under doubled atmospheric CO_2 (Lemon, 1983). Plant competition undoubtedly eliminates many species from most sites. In making these projections the possibility of competition from angiosperms, which may be severe in conditions of doubled atmospheric CO_2, is completely ignored.

CURRENT SPECIES DISTRIBUTION

Current distribution of ponderosa pine, Douglas-fir, western larch, western hemlock, lodgepole pine, and Engelmann spruce are shown in Little (1971). It was difficult to relate these maps to the forecast climate changes. To do so, Table 1, based on the amount of overlap between the species distribution maps of Little (1971) and the National Climate Center's climate divisions (Gibbs and Hoffman, this volume), was prepared (Figure 2). A listing of historical climate data (1931 to 1980) was used to develop the table. More accurate temperature, elevation, and precipitation ranges are available from a number of sources (Fowells, 1965; Barrett, 1980). However, such data are difficult to interpret at the level of climate divisions and have other inherent problems.

PROJECTED CLIMATE CHANGES

The effects on global regional climate caused by a doubling of

Table 1
Upper and Lower Limits to Climate Data, 1931-1980

(from the National Climate Center's climate divisions that contain tree species in the percent covered given below)

	Percent of division covered	Western hemlock	Ponderosa pine	Douglas-fir	Lodgepole pine	Western larch	Englemann spruce
Maximum mean monthly temperature	75-100%	19oC	20oC	20oC	19oC	19oC	19oC
	50- 74	19oC	23oC	20oC	19oC	19oC	19oC
	25- 49	19oC	23oC	20oC	21oC	19oC	19oC
	1- 24	20oC	29oC	29oC	25oC	21oC	27oC
Minimum annual precipitation	75-100%	59cm	45cm	23cm	39cm	52cm	231cm
	50- 74	59cm	31cm	23cm	23cm	45cm	52cm
	25- 49	59cm	31cm	23cm	23cm	45cm	29cm
	1- 24	48cm	20cm	23cm	23cm	31cm	29cm
Minimum monthly precipitation	75-100%	18mm	4mm	4mm	10mm	20mm	10mm
	50- 74	17mm	3mm	4mm	10mm	16mm	10mm
	25- 49	13mm	3mm	4mm	7mm	10mm	10mm
	1- 24	4mm	1mm	1mm	1mm	10mm	4mm
Maximum mean January temperature	75-100%	6oC	6oC	6oC	4oC	-3oC	1oC
	50- 74	6oC	6oC	6oC	6oC	-3oC	1oC
	25- 49	6oC	6oC	6oC	6oC	-3oC	1oC
	1- 24	6oC	11oC	9oC	6oC	-3oC	6oC

Figure 2

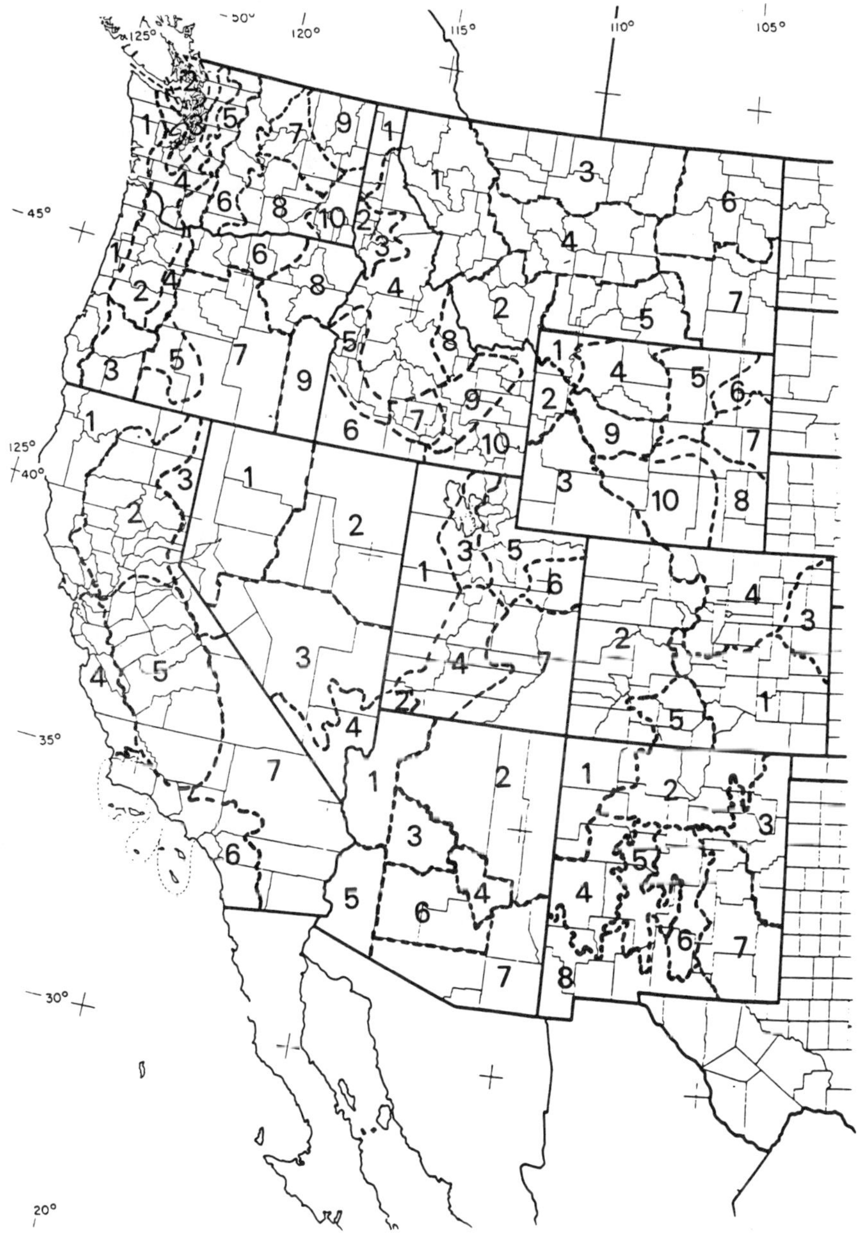

National Climate Center's climate divisions for the western United States.

CO_2 in this planet's atmosphere are projected to be as follows (Gibbs and Hoffman, this volume).

Temperature

Changes anticipated in mean annual temperature with doubled atmospheric CO_2 are presented in Figure 3. The increases range from less than 2°C (3.6°F) (southern California coast) to slightly above 6°C (11°F) in Arizona, New Mexico, southern Colorado, and southern Utah. The warming generally increases from west to east and from north to south.

Because increased levels of CO_2 compensate for increased temperatures, and because temperatures do not rise much more than 6°C (11°F), no direct negative effect of rising temperature on growth rates is expected to be caused by a decrease in photosynthesis relative to respiration. In some cases, the increase in mean monthly temperature in the ICF scenario (this volume) may push many exosed soil-surface temperatures above 54°C (130°F), a temperature at which seedlings are killed by acute heat girdling (Barrett, 1980). However, the most important factors for the buildup of lethal soil-surface temperatures are soil-surface aspect, texture, moisture, and color (Vaartaja, 1954). With the proper combination of factors, soil-surface temperature can be 50°C (90°F) higher than air temperature (Geiger, 1966). Currently, high-temperature girdling of tree seedlings does not significantly limit the distribution of tree species on the scale of the accompanying maps. Because shaded microsites are common, it is assumed that, with doubled CO_2, high-temperature girdling of trees will remain insignificant.

Changes in temperature appear relatively minor compared with the seasonal amplitude of temperature on inland sites (Figure 4). Thus, winter temperatures on these sites remain sufficiently low to meet the chilling requirements of the six conifers. However, in the coastal zones winter temperatures rise above 9°C (48°F), the assumed requirement for chilling in some species (Figure 5).

Daubenmire (1974) noted a general 0.5°C (0.9°F) temperature rise per 100-m (350-foot) decrease in elevation. Although lower altitude limits for trees will not be pushed up because of increased temperature effects on carbon balance, the upper limits certainly will be. A rise in the upper elevation limit of species is projected to be 250 to 1200 m (800 to 3900 feet) if CO_2 in the atmosphere doubles. Therefore, reductions of subalpine meadows and subalpine tree species in most mountain ranges are anticipated. Frost pockets will disappear in many areas, in part because the denser atmo-

Figure 3

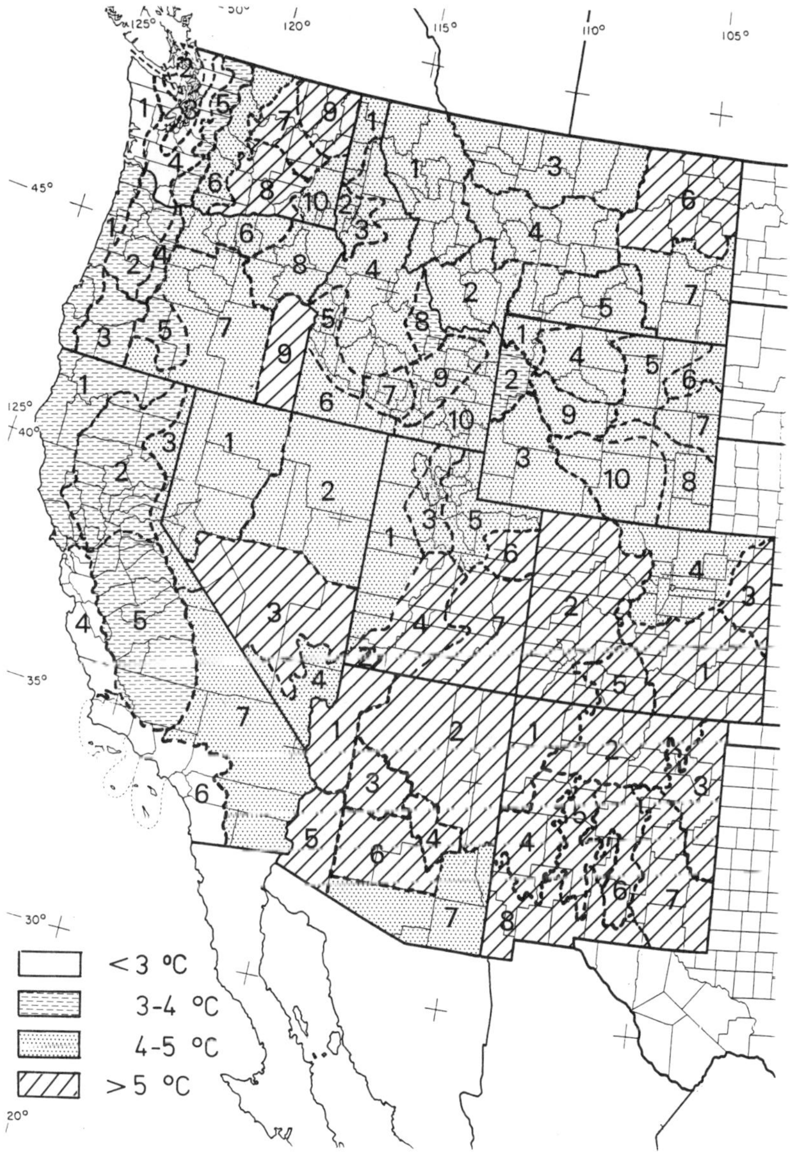

Projected increases in mean annual temperatures for climate divisions of the western United States. These values are for the doubled atmospheric CO_2 climate scenario minus the averages for the period from 1931 to 1980. Data were provided by Gibbs and Hoffman (this volume).

Figure 4

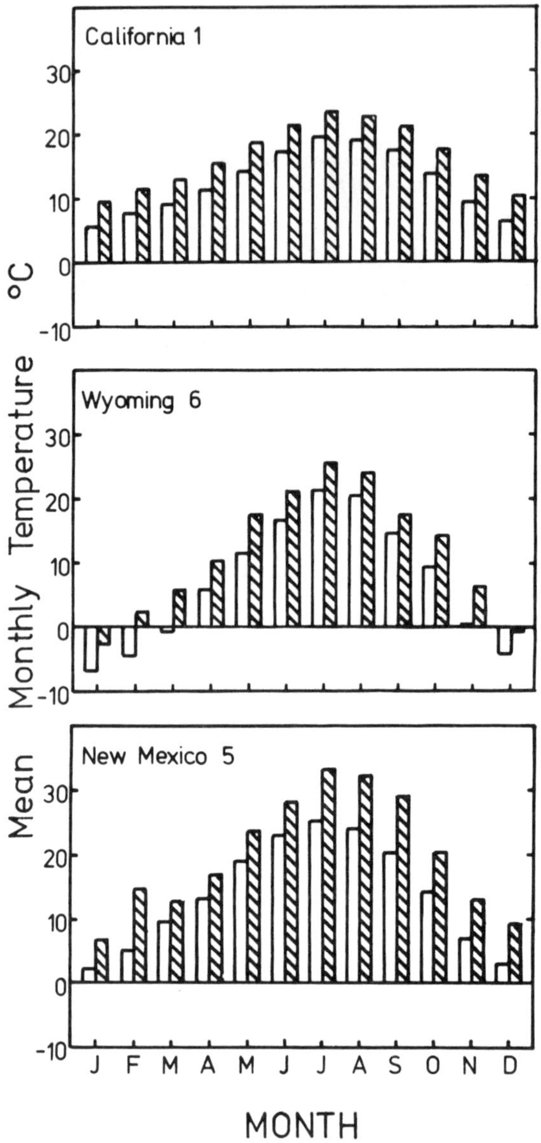

Monthly mean temperatures for the period from 1931 to 1980 (open bars) and for the doubled-CO_2 climate scenario (hatched bars). These three histograms represent the range occurring within the climate divisions of the western United States.

Figure 5

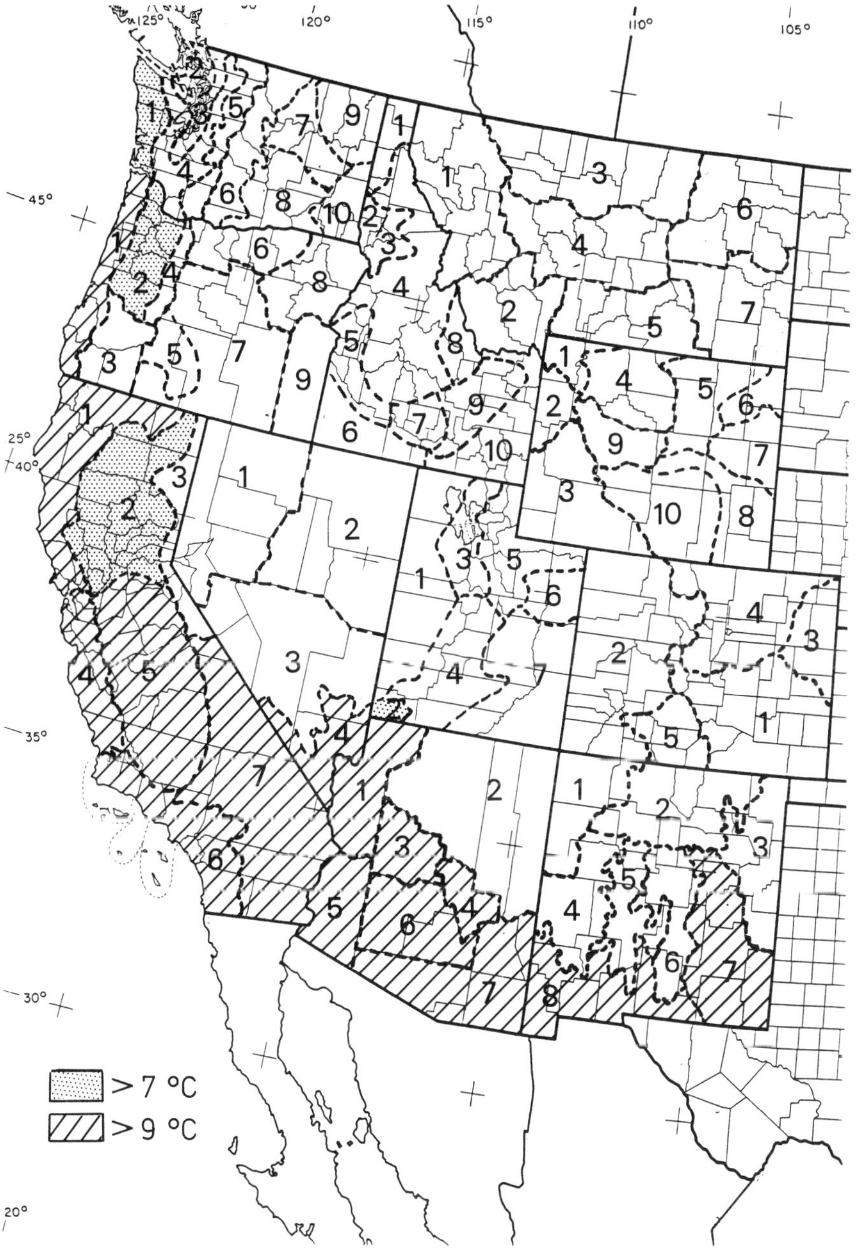

Maximum winter chilling in the doubled-CO_2 climate scenario. Shading indicates minimum mean temperature for the coldest month.

spheric CO_2 intercepts more infrared radiation and reduces differences in temperature between day and night.

Water Stress

Next, increased evaporative demand minus precipitation was estimated, based on the climate scenario provided by ICF (Figure 6) (Gibbs and Hoffman, this volume), for each climate division by:

1) calculating saturated vapor pressure of historical mean annual temperature;

2) calculating saturated vapor pressure of the doubled-CO_2 mean annual temperature;

3) dividing 2 by 1 to get the relative increase in evaporative demand (expressed as a percent increase; that is, 1.28 = 28% increase); and

4) subtracting the percent increase in precipitation as predicted by the climate scenario from the results in step 3. For example, for the Washington division, 1:29 - 30 = -1% increase; for the Arizona division, 2:39 - (-7) = 46% increase.

This method is necessarily crude but represents a summary consistent with data and time limitations. It should be noted that, because of their small dimensions, conifer needles are closely coupled to air temperature and that transpiration of conifer forests depends strongly on vapor pressure deficit (Jarvis and Stewart, 1979). Evaporation from leaves at air temperature should increase in roughly the same proportion as the increase in saturation vapor pressure (Geiger, 1966).

Precipitation during the growing season is critical for natural regeneration (Barrett, 1980). Therefore, a second map of the change in site water balance was developed (Figure 7). In this case, the single month from June through September with the largest decrease in precipitation (or, in a few cases, with the smallest increase) was subjected to the same calculations used for Figure 6. Eleven climate divisions have a decrease in site water balance under 25% (Idaho, 7, 8, 9, and 10; Wyoming, 1, 2, 3, and 4; Montana, 2; and Utah, 1 and 5; Figure 7). These are the only places for which this scenario projects a definite favorable increase in plant-water balance in the summer. For most of Oregon, Washington, Idaho, and Montana plant-water balance will be about the same (25 to 50% increase in evaporation over precipitation). For eastern Wyoming, eastern Colorado, southeastern Utah, Arizona, New Mexico, and California, the doubled CO_2 will not fully compensate for the less favorable water balance of the projection. That is, these areas should be both physically and physiologically

Figure 6

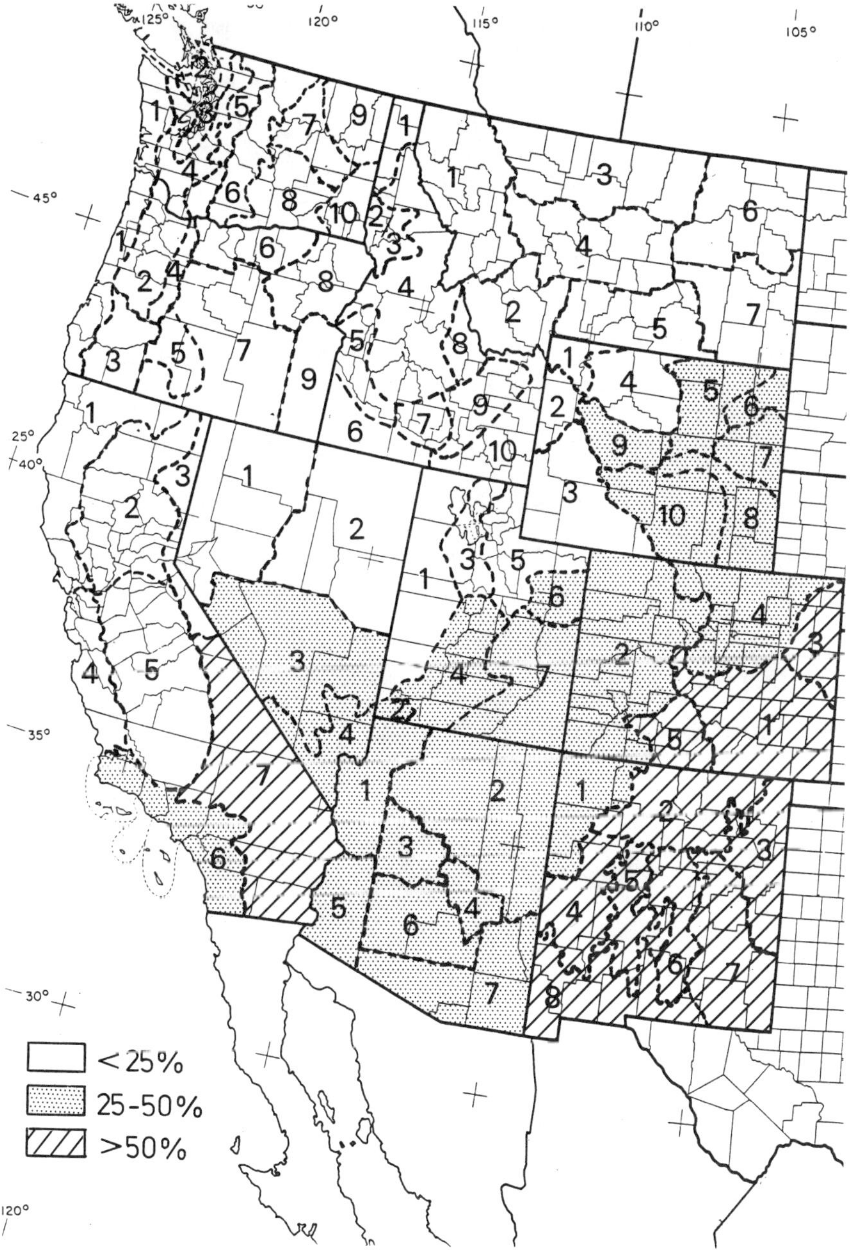

Index of the relative decrease in site water balance (see text for calculation) based on mean annual temperature and mean annual precipitation.

Figure 7

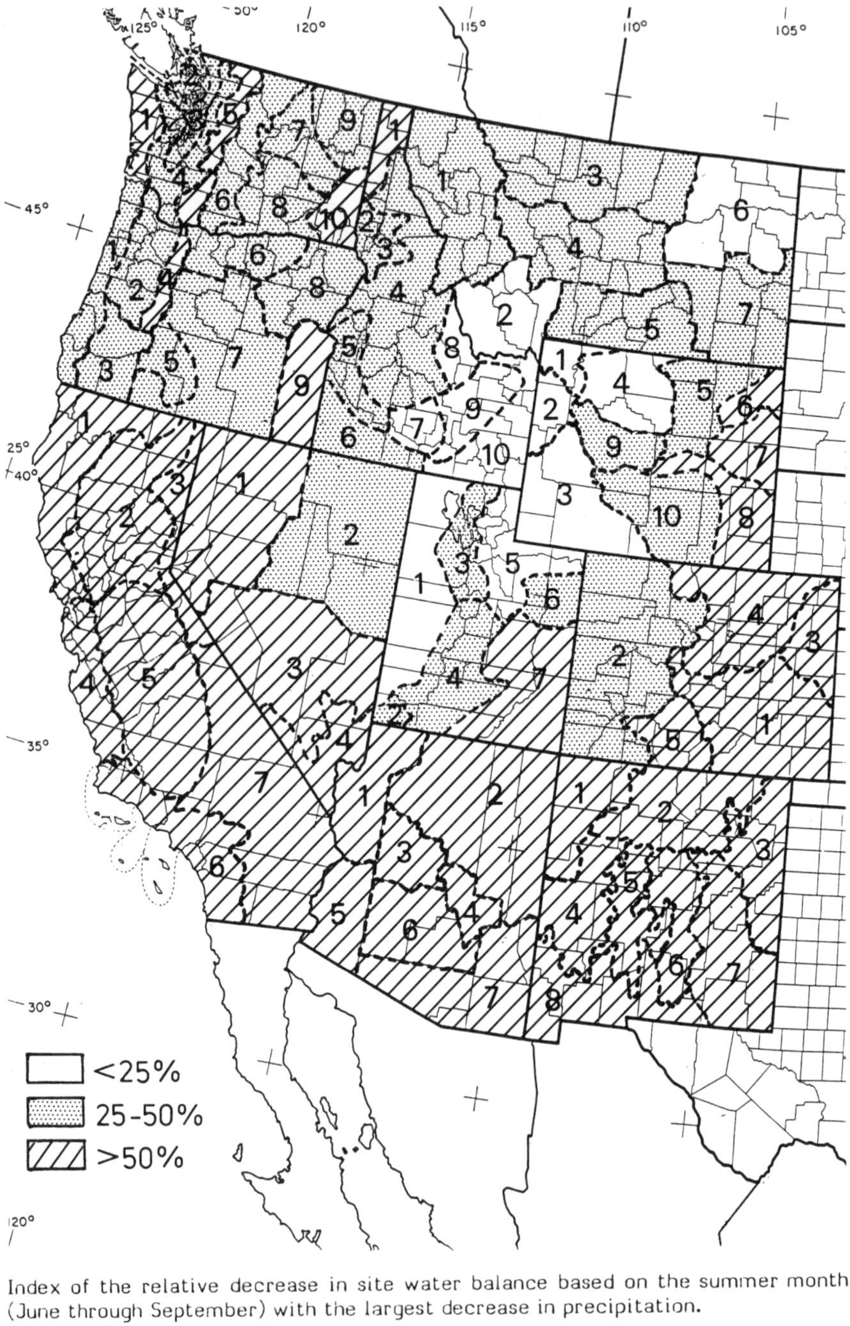

Index of the relative decrease in site water balance based on the summer month (June through September) with the largest decrease in precipitation.

drier. Some of these numbers may be misleading. For example, on the coast of Washington State (Washington division 1; Figure 7), high summer precipitation presently predominates, so even a large absolute change in precipitation leaves sufficient precipitation for tree growth.

Spring precipitation is also important for regeneration (Barrett, 1980). Therefore, the above analysis was repeated for the months of March through May (Figure 8). Shoot growth and germination currently starts in April. With increases in temperature (Figure 4), spring is projected to arrive one month earlier provided chilling requirements are met. Therefore the month of March was included. However, at high altitudes spring may not arrive until late July. In the doubled-CO_2 scenario, site water balance is higher in the spring than in the summer in California (Figures 7 and 8). For the whole eastern slope of the Rocky Mountains, from Canada to Mexico, there is a physiologically significant decrease in site water balance in spring (Figure 8).

PROJECTED SPECIES DISTRIBUTION

In these genetically heterogeneous conifers there may be considerable potential for rapid genetic change. Therefore, if the genotype exists within the general area (that is, on the West Coast or in the Rocky Mountains), the species was assumed to survive even if the local genotypes adapted poorly to the new local climate. Obviously, individual trees, with their life spans of hundreds of years, have endured large variations in weather in the past (Campbell 1978).

Ponderosa Pine

Large areas of ponderosa pine occur in every state in the West except Nevada (Figure 9). High-quality wood has made it a top commercial species. Ponderosa pine is able to endure long summer droughts provided the seedlings get well established in a moist spring (Fowells, 1965). Therefore, in this scenario ponderosa pine expands, at the expense of its competitors, in the forest areas of California where springs are physiologically as dry as now but where summers will be drier (Figures 7 and 8). No information on chilling requirements of ponderosa pine is available. However, the lower elevation limit of ponderosa pine is projected to be pushed upslope on the coast of California, because the species does not occur in any climate division in which mean monthly temperatures always stay above 9°C (48°F) (Table 1). The upper elevation limit

Figure 8

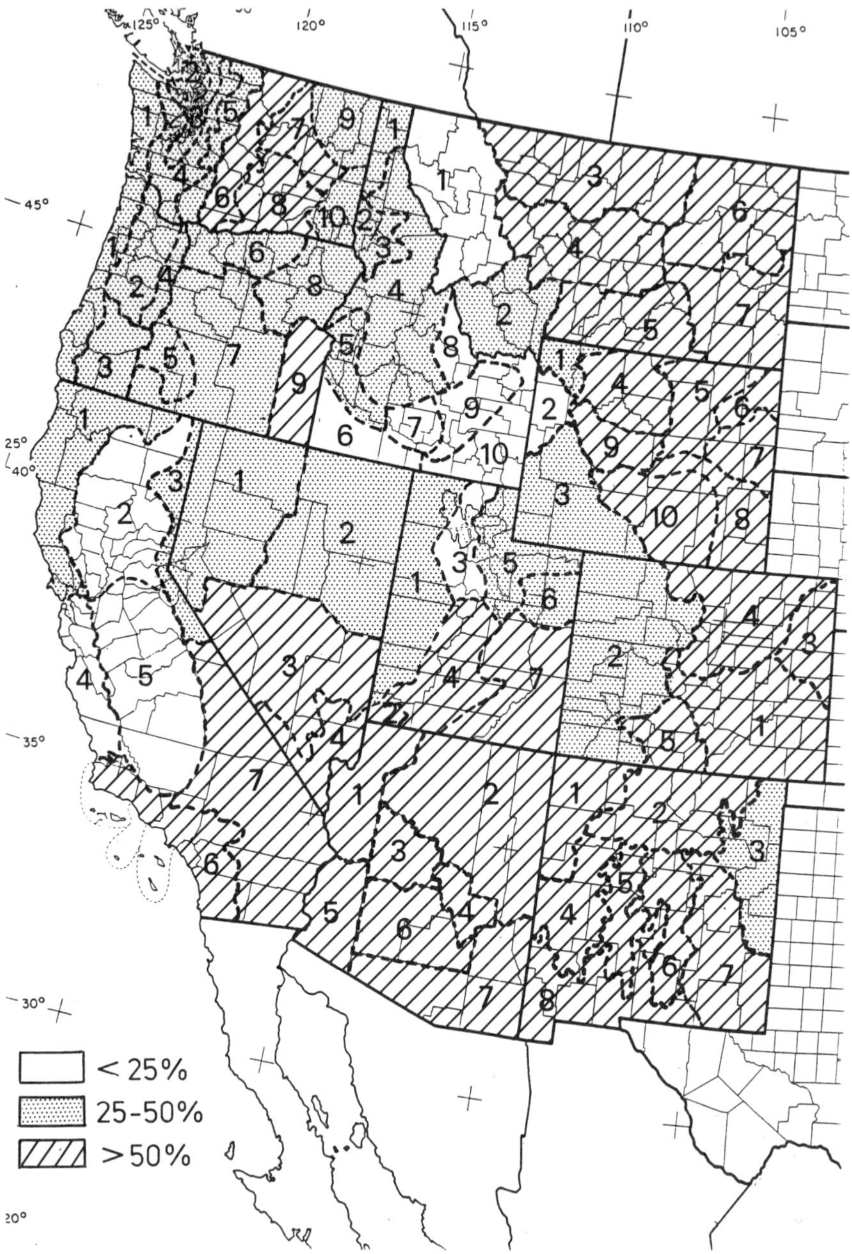

Index of the relative decrease in site water balance based on the spring month (March through May) with the largest decrease in precipitation.

Figure 9

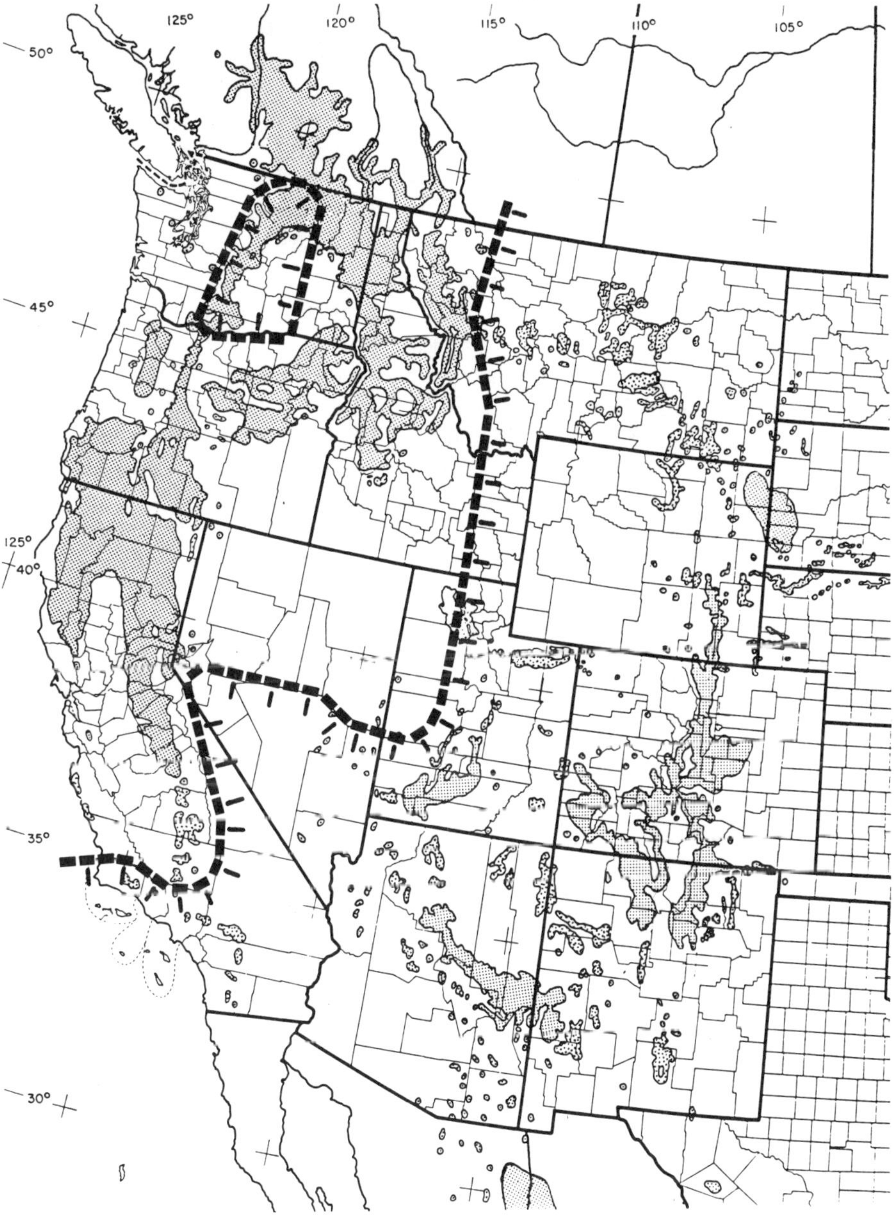

The current natural distribution of ponderosa pine (Little, 1971). Hatching is directed toward the zones of decreasing acreage, as projected in the doubled-CO_2 scenario.

of ponderosa pine is strongly affected by its high heat requirements (Daubenmire, 1943; Daniel, 1980). The increased temperatures of the doubled-CO_2 scenario are projected to allow the upper elevation limit of ponderosa pine to move upslope by 800 m (2600 feet) in California, between 500 and 1100 m (1600 and 3600 feet) in Washington and Oregon, by 800 to 1000 m (2600 to 3300 feet) in Montana and Idaho, and between 900 and 1300 m (3000 and 4300 feet) in the middle and southern Rocky Mountains. However, this expansion occurs only in the absence of competition. Competitive exclusion by more mesic species, especially Douglas-fir, is projected to limit the upper range of ponderosa pine considerably below these upper limits where summer drought is not severe.

The reduction in spring site water balance in eastern Washington (divisions 6, 7, and 8) and on the eastern slope of the Rocky Mountains from Canada to New Mexico forces ponderosa pine upslope. In Oregon's Cascade Mountains, ponderosa pine expands considerably as a result of a more severe summer drought. A major decrease in lower elevation ponderosa pine forests on the eastern slope of the Rocky Mountains from Canada to Mexico is projected. The reduction in spring water balance also forces its competitors upslope in these areas, so its upper elevation limit moves upslope as well.

In summary, the area covered by ponderosa pine is projected to increase in California and Oregon's Cascade Mountains. It is projected to decrease on the eastern slope of the Washington Cascade and the Rocky Mountains, as well as in Arizona and New Mexico (Figure 9).

Douglas-Fir

Douglas-fir is the dominant commercial forest species of the Pacific Northwest and of the northern Rocky Mountains. It is also an important commercial species in most other regions of the West (Barrett, 1980). The upper elevation limit of coastal Douglas-fir is determined by the winter snowpack, which physically damages seedlings (Scott, 1980), and cold soil temperatures, which can induce water stress (Teskey, 1982). The rise in temperature in the doubled-CO_2 scenario is projected to move the lower elevation limit of the continuous winter snowpack in Washington and Oregon 700 to 800 m (2300 to 2600 feet) upslope. In Washington State this would raise the altitude of the winter snowpack from the current 800 m (2600 feet) to 1500 m (4900 feet). Therefore, the upper elevation limit of Douglas-fir on the west side of the Cascade Mountains is expected to move dramatically upslope (Figure 10). At

Figure 10

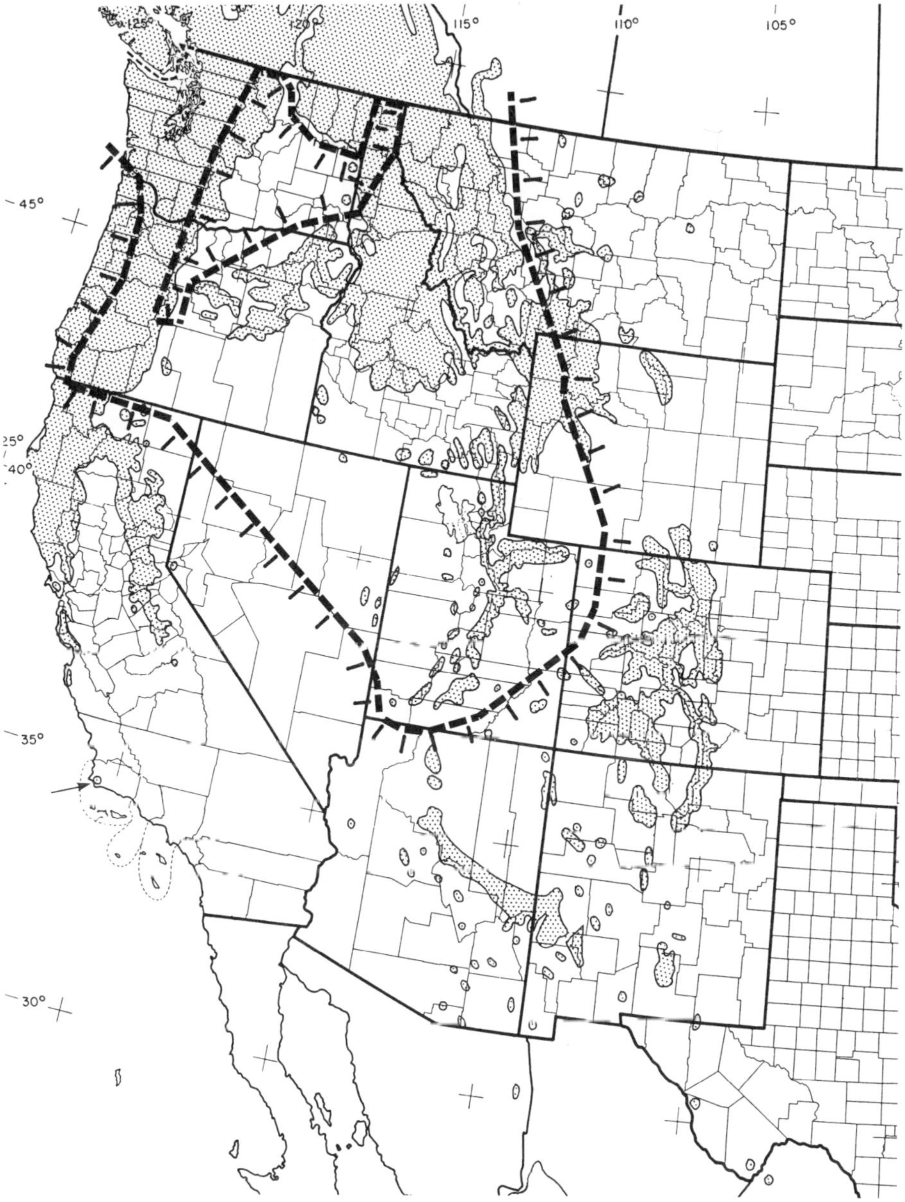

The current natural distribution of Douglas-fir (Little, 1971). Hatching is directed toward the zones of decreasing acreage, as projected in the doubled-CO_2 scenario.

the same time the doubled atmospheric CO_2 favorably affects carbon balance and allows Douglas-fir to maintain its current lower elevation limit at sea level.

Rising winter temperatures do not meet the assumed chilling requirements for Douglas-fir at lower elevations in coastal areas of Oregon and California and the west slope of the Sierra in California. Douglas-fir is projected to remain in these climate zones, but it will move upslope. In California the long summer drought will cause Douglas-fir to retreat to cooler, wetter sites. In the northern part of the Rocky Mountains, Douglas-fir is projected to maintain its lower elevation limits, as a result of CO_2 compensation for lower site water, and to expand upslope by 900 to 1300 m (3000 to 4300 feet). In the relatively wet zone centered around southern Idaho (Figures 7 and 8), Douglas-fir is projected to expand downslope.

In summary, Douglas-fir is projected to expand or maintain its present position over most of its commercially important range (Figure 10). However, the species decreases in area in California, coastal Oregon, eastern Washington, on the eastern slope of the Rocky Mountains, and in Arizona and New Mexico.

Western Hemlock

Outside the Pacific Northwest western hemlock is of minor commercial importance. However, within the Pacific Northwest its total volume matches that of Douglas-fir (Barrett, 1980). The upper elevation limit of western hemlock is also affected by the depth of the winter snowpack but less than Douglas-fir (Scott, 1980). Western hemlock also is potentially limited by water stress induced by cold soil temperatures (Teskey, 1982). Therefore, along with Douglas-fir, western hemlock is expected to expand upslope 700 to 1100 m (2300 to 3600 feet) in the Cascade and northern Rocky Mountains.

Compared with Douglas-fir, western hemlock has a lower chilling requirement (Nelson and Lavender, 1979). Therefore, western hemlock is projected to be more dominant in coastal Oregon. Although the summer drought of the Washington coast is severe in terms of percent change, the minimum average monthly rainfall is 41 mm (1.6 in.) in this scenario, well above that required for western hemlock (Table 1). However, the summer drought on the western slope of the Washington and Oregon Cascade Mountains is severe, and the minimum average monthly precipitation is 15 and 0.0 mm (0.59 and 0.0 in.), respectively. Hence, even with the compensatory effect of doubled atmospheric CO_2, western hemlock is expected to be limited to cool, wet sites or even eliminated from the Oregon Cascade Mountains (Figure 11). A reduction in the area of western hemlock in northern Idaho is also projected because of drought (Figure 11).

In summary, western hemlock is projected to expand in area in western Washington along the Oregon coast and in some areas of the northern Rocky Mountains. Western hemlock is projected to decrease in importance in Oregon east of the Willamette Valley and in northern Idaho.

Western Larch

Western larch is restricted largely to the northern Rocky Mountain region (Figure 12). However, within this region the annual volume cut is second only to Douglas-fir (Adams, 1980). Initial height growth of western larch is larger than any of its competitors. On sites prone to frost, lodgepole and ponderosa pine will out-compete frost-susceptible western larch. In addition, western larch is relatively sensitive to warm temperatures (perhaps as they affect transpiration) and thus is limited largely to cool valley sites and northern and eastern slopes. Because western larch is genetically programmed for fast leader growth, it is expected to respond more to doubled atmospheric CO_2 than Douglas-fir, western hemlock, ponderosa pine, or Engelmann spruce on the better sites. Western larch regenerates well after fire, and if increased summer drought induces higher fire frequency, the species is projected to be highly favored in this zone. Western larch is projected to maintain its present lower elevation limit as a result of CO_2 compensating for higher temperatures. In addition, it is projected to expand further upslope by 800 to 1000 m (2600 to 3300 feet) to out-compete the slower growing lodgepole pine and Engelmann spruce on disturbed sites. The species also is seen expanding into areas where it is currently excluded because of frosts. Western larch growing in the eastern Washington and Oregon Cascade Mountains will experience significantly more severe summer drought and will also experience a more severe spring drought in the Washington Cascade Mountains. Therefore, in the Cascade Mountain region, western larch area declines as it is forced upslope to a cooler, moister climate (Figure 12).

Lodgepole Pine

Lodgepole pine is the third most abundant timber species in the West and has a greater area and volume in the Rocky Mountains

Figure 11

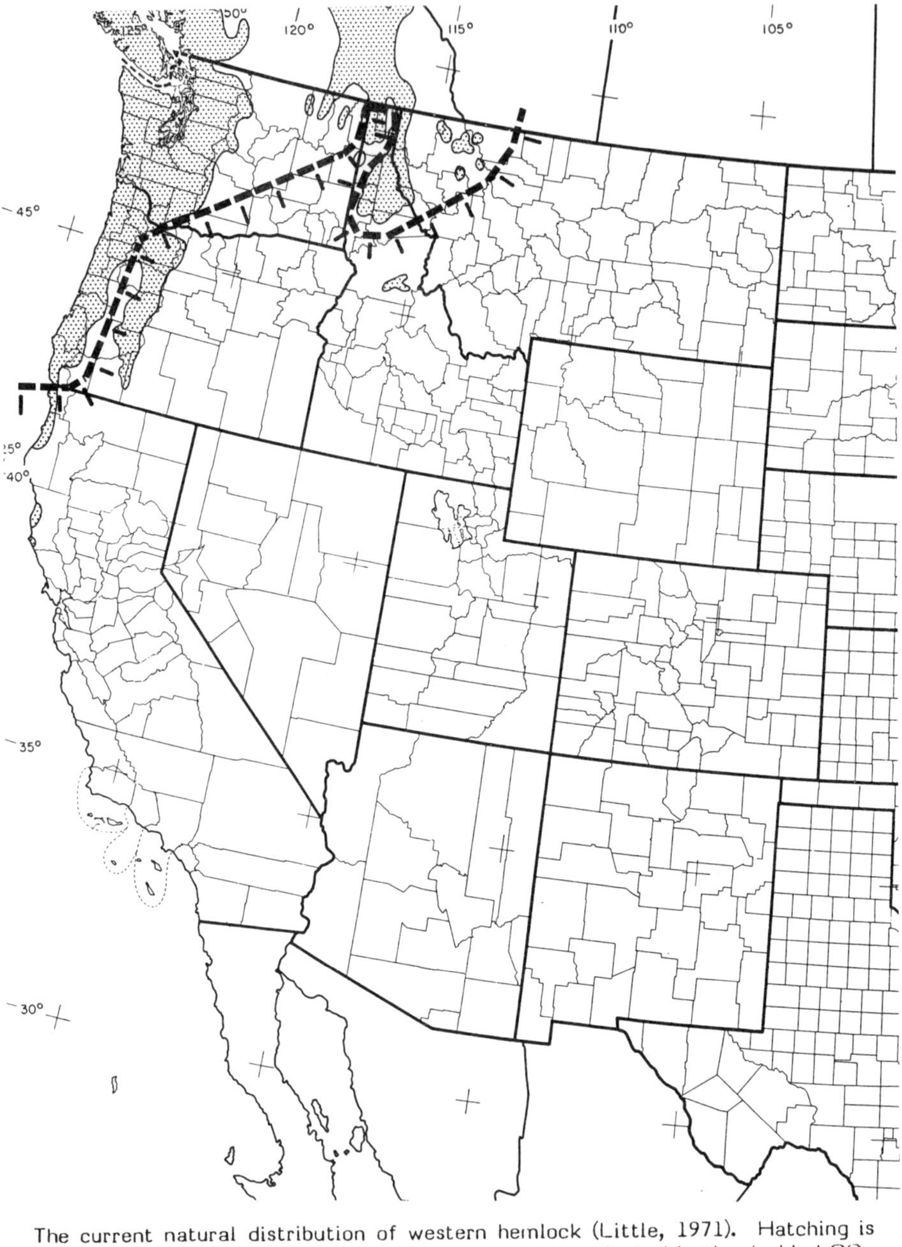

The current natural distribution of western hemlock (Little, 1971). Hatching is directed toward the zone of decreasing acreage, as projected in the doubled-CO_2 scenario.

Figure 12

The current natural distribution of western larch (Little, 1971). Hatching is directed toward the zone of decreasing acreage, as projected in the doubled-CO_2 scenario.

Figure 13

The current natural distribution of lodgepole pine (Little, 1971). No significant net decrease in acreage is projected for lodgepole pine.

than Douglas-fir (Barrett, 1980). Most of its range is not expected to be negatively affected (Figure 13). Because the tree is especially favored by fire, lodgepole pine is projected to regenerate much more aggressively in some areas, as the frequency of wildfires is assumed to increase. Lodgepole pine often occurs on quite demanding sites where timing of germination and shoot growth must be precise so the plant is hardened for summer drought or winter conditions (Cleary et al., 1978). Problems with local genotypes not well timed to the new climate of the doubled-CO_2 scenario may be serious. However, under equilibrium conditions, the gene pool is assumed to be large enough that such effects will be insignificant. Increasing temperatures and decreasing day-night temperature differences eliminate many frost pockets and allow some upper elevation flats to be occupied by ponderosa pine on the drier sites and Douglas-fir, western larch, or Engelmann spruce on the cooler, wetter sites. In the physiologically wet zone surrounding southern Idaho, lodgepole pine is projected to decrease in importance because of increased competition from Douglas-fir; however, it should also move downslope into areas currently too dry.

In summary, over most of the West the area of lodgepole pine is projected to expand and shrink depending on local conditions. The net effect is no major change in the amount of area the species occupies (Figure 13).

Engelmann Spruce

Engelmann spruce is a high-altitude, shade-tolerant species characterized by slow early-juvenile growth (Barrett, 1980). Root growth is so slow that the tree requires several years of favorable weather to get its roots to a depth sufficient to maintain it through droughts. Because this spruce is genetically programmed to grow slowly, it is projected to respond less vigorously to doubled atmospheric CO_2 than Douglas-fir, lodgepole pine, western hemlock, and western larch on better sites. However, the success of shade-tolerant species does not necessarily depend on fast growth. While Engelmann spruce is projected to decrease in importance on disturbed sites, it maintains itself in stand competition.

The upper elevation limit to Engelmann spruce is often the tree line, where high winds, deep snowpack, and a complete lack of soil (Daubenmire, 1943) may act as factors that cannot be compensated for by increased atmospheric CO_2. At present Engelmann spruce occurs above the zone of maximum precipitation on the mountain slopes (Daubenmire, 1943). Because warm air holds more water, it is possible for precipitation or snowpack to substantially increase in and above the Engelmann spruce zone as precipitation decreases at lower elevations. As a result of increased snowpack, the upper elevation limit of Engelmann spruce is not projected to move uniformly upslope as much as is indicated by increases in temperature. Engelmann spruce is projected to invade frost pockets currently occupied by the more frost-tolerant lodgepole pine.

In summary, the acreage of Engelmann spruce is expected to decline over its whole range because of strong competition on disturbed open sites at its lower elevation limit and increased snowpack and lack of soil at higher elevations (Figure 14).

SUMMARY

In this scenario the Rocky Mountains of Idaho and Montana are affected favorably by rising atmospheric CO_2 in that the valuable commercial tree species, Douglas-fir, ponderosa pine, and western larch, will expand their ranges upslope by perhaps 700 to 1100 m (2300 to 3600 feet) and still maintain their lower elevation limits. In parts of southern Idaho, southwestern Montana, Utah, and Wyoming, downslope expansions of Douglas-fir are predicted. In this physiologically moist zone, lodgepole pine decreases in importance at the upper half of its elevation range. Ponderosa pine and western larch do not currently exist there.

Western Washington also is affected favorably by a projected increase in Douglas-fir. In the Cascade Mountains, an expanded importance of Douglas-fir relative to western hemlock is projected because of a more intense summer drought. On the western slope of Oregon's Cascade Mountain range, an expansion of ponderosa pine up from the Willamette Valley is projected. On the eastern slope of the Cascade Mountains of Washington, natural regeneration of all species is projected to move upslope due to dry springs.

In California, ponderosa pine is projected to expand as other species retreat, a result of more intense summer drought. Douglas-fir is eliminated from the lowlands of California and Oregon coastal divisions because chilling requirements are not met for both shoot growth and seed germination. Lodgepole pine is projected to expand on some sites and to be reduced in others as night frosts become less severe.

The southern and eastern Rocky Mountains are negatively affected in the doubled-CO_2 scenario. The lower elevation limits of commercial species are expected to move upslope due to increased water stress. While the upper limit is projected to also move up-

Figure 14

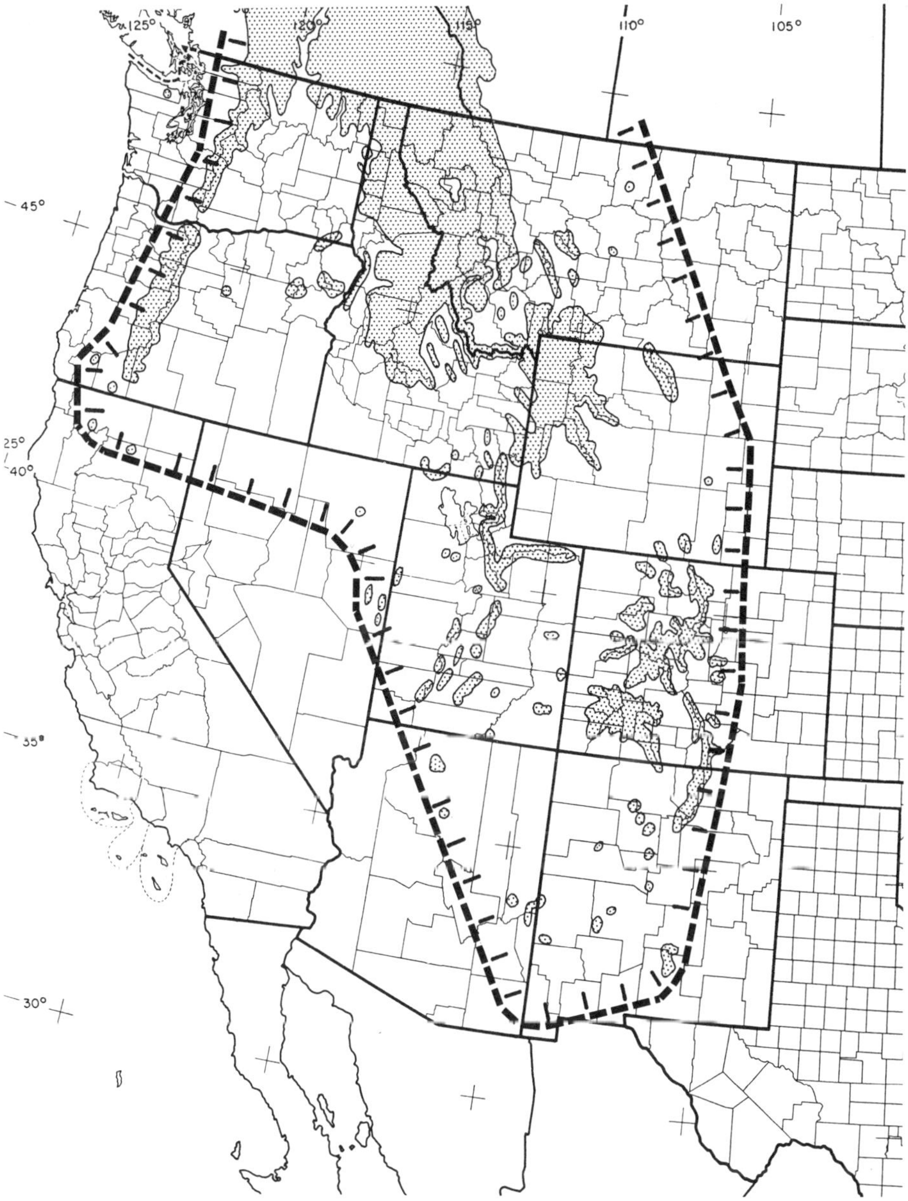

The current natural distribution of Engelmann spruce (Little, 1971). Engelmann spruce is projected to decrease over its entire range.

Figure 15

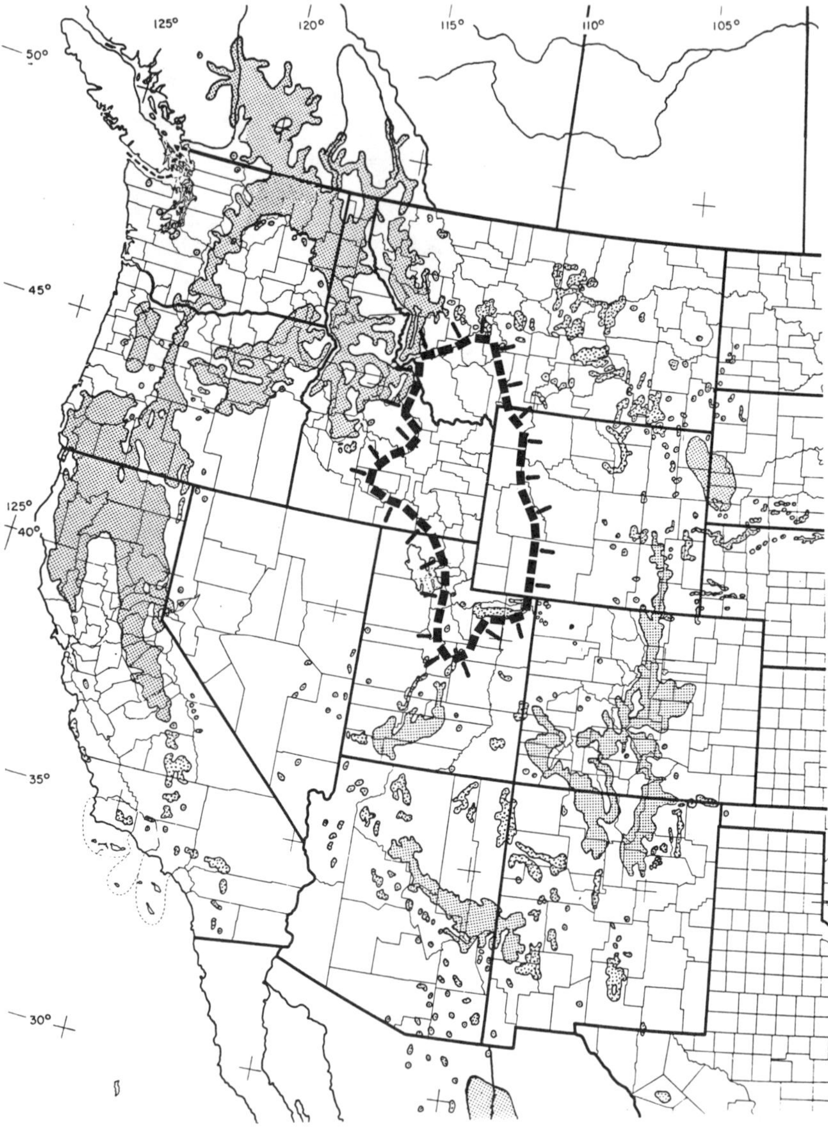

The current natural distribution of ponderosa pine (Little, 1971). Hatching is directed toward the zones of decreasing acreage, as projected if a doubled atmospheric CO_2 only compensates for a 25% decrease in site water balance.

hill, acreages will decrease. Species will be eliminated from the tops of some mountains.

A SECOND SCENARIO

Perhaps the most important assumption about plant response deals with the ability of doubled atmospheric CO_2 to compensate for a 25 to 50% decrease in site water balance in conifers. For example, if it is assumed, reasonably, that doubled CO_2 content will only compensate for up to a 25% decrease in site water balance, dramatic shifts in the projected changes occur. With this lower assumed efficiency, almost the whole West is projected to become physiologically drier (Figure 15). Only in a small zone centered around southern Idaho will sites not become significantly more droughty. Thus, in this second scheme major shrinkages of forest zones are projected as species move upslope almost everywhere. This emphasizes the extreme importance of being able to precisely define how the major forest species will respond to increased atmospheric CO_2 under water stress.

A second important assumption concerns the chilling requirement of these species. If Douglas-fir, or any other species, does not receive sufficient chilling, when mean monthly temperatures do not drop below 7°C (45°F) instead of the 9°C (48°F) assumed above, Douglas-fir will be eliminated from lower elevation sites along the entire West Coast of the United States, in the Willamette Valley of Oregon, and in the Sierra ranges of California.

REFERENCES

Adams, D. L. 1980. The Northern Rocky Mountain region, p. 341-389. *In* J. W. Barrett (ed.), *Regional Silviculture of the United States*. John Wiley & Sons, Inc., New York.

Allen, R. M., and W. H. D. McGregor. 1962. Seedling growth of three southern pine species under long and short days. *Silv. Genet.* 11, 43-45.

Barrett, J. W. 1980. *Regional Silviculture of the United States.* John Wiley & Sons, Inc., New York.

Berry, J. A., and J. K. Raison. 1981. Responses of macrophytes to temperature. pp. 277-338. *In* O. L. Lange, P. S. Nobel, C. B. Osmond, and H. Ziegler (ed.), *Physiological Plant Ecology, vol. I.* Springer-Verlag, New York.

Campbell, R. K. 1978. Regulation of bud burst timing by temperature and photoregime during dormancy. *In* C. A. Hollis (ed.), *Proceedings of the Fifth North American Forest Biology Work-*

shop.

Cleary, B. D., R. D. Greaves, and R. K. Hermann. 1978. *Regenerating Oregon's Forests.* Oregon State University Press, Corvallis.

Copes, D. L. 1983. Failure of grafted Douglas-fir planted at Monterey, California. *Tree Planters Notes* 34: 9-10.

Daniel, T.W. 1980. The Middle and Southern Rocky Mountain region, p. 277-340. *In* J. W. Barrett (ed.), *Regional Silviculture of the United States.* John Wiley & Sons, Inc., New York.

Daubenmire, R. F. 1943. Vegetational zonation in the Rocky Mountains. *Bot. Rev.* 6, 325-393.

Daubenmire, R. F. 1974. *Plants and Environment.* John Wiley & Sons, Inc., New York.

Ehleringer, J., and O. Bjorkman. 1977. Quantum yield for CO_2 uptake of C_3 and C_4 plants. Dependence on temperature, CO_2 and O_2 concentration. *Plant Physiol.* 59, 86-90.

Fowells, H. A. 1965. *Silvics of Forest Trees of the United States.* USDA Forest Service Agriculture Handbook 271. U.S. Department of Agriculture, Washington, D.C.

Geiger, R. 1966. *The Climate Near the Ground.* Harvard University Press, Cambridge, Mass.

Jarvis, P. G., and J. Stewart. 1979. Evaporation of water from plantation forest, p. 327-349. *In* E. D. Ford, D. C. Malcolm, and J. Atterson (ed.), *The Ecology of Even-Aged Forest Plantations.* Institute of Terrestrial Ecology Publishers, Cambridge, England.

Jarvis, P. G., and J. I. L. Morison. 1981. The control of transpiration and photosynthesis by the stomata, p. 247-279. *In* P. G. Jarvis and T. A. Mansfield (ed.), *Stomatal Physiology.* Cambridge University Press, Cambridge, England.

Jarvis, P. G., and J. W. Leverenz. 1983. Productivity of temperate, deciduous and evergreen forest, p. 233-280. *In* O. L. Lange, P. S. Nobel, C. B. Osmond, and H. Ziegler (ed.), *Physiological Plant Ecology,* vol. IV. Springer-Verlag, Berlin.

Lavender, D. P. 1981. *Environment and Shoot Growth of Woody Plants.* Forest Research Laboratory Publication, School of Forestry, Oregon State University, Corvallis.

Lavender, D. P. 1984. Greenhouse effect. *J. of Forestry* 82, 211.

Lemon, E. R. (ed.). 1983. CO_2 and plants. *In The Response of Plants to Rising Levels of Atmospheric Carbon Dioxide.* Westview Press, Boulder, Colo.

Little, E. L., Jr. 1971. *Atlas of United States Trees, Vol. 1. Conifers and Important Hardwoods.* U.S. Government Printing Office, Washington, D.C.

Morison, J. I. L., and R. H. Gifford. 1984. Plant growth and water use with limited water supply in high CO_2 concentrations. 1.

Leaf area, water use and transpiration. *Aust. J. Plant Physiol.* (in press).

Nelson, E. A. and D. P. Lavender. 1979. The chilling requirement for western hemlock seedlings. *Forest Sci.* 25, 485-490.

Purohit, A. N. and E. B. Tregunna.. 1976. Effects of carbon dioxide on the growth of Douglas-fir seedlings. *Indian J. Plant Physiol.* 19, 164-170.

Scott, D. R. M. 1980. The Pacific Northwest region, p. 447-493. *In* J. W. Barrett (ed.), *Regional Silviculture of the United States.* John Wiley & Sons, Inc, New York.

Teskey, R. O. 1982. Acclimation of *Abies amabilis* to water and temperature stress in a natural environment. Ph.D. Thesis, University of Washington, Seattle.

Tinus, R. W. 1972. CO_2 enriched atmosphere speeds growth of ponderosa pine and blue spruce seedlings. *Tree Planters Notes* 23(1), 12-15.

Vaartaja, O. 1954. Temperature and evaporation at and near ground level on certain forest sites. *Can. J. of Bot.* 32, 760-783.

EFFECT OF RISING CARBON DIOXIDE AND POTENTIAL CLIMATE CHANGE ON LOBLOLLY PINE DISTRIBUTION, GROWTH, SURVIVAL, AND PRODUCTIVITY

by W. Frank Miller, Philip M. Dougherty, and George L. Switzer

ABSTRACT

An appreciable northward and northeastward shift in the range of loblolly pine is forecast by the year 2080. The projected southern boundary of the range may extend from central Louisiana through central Mississippi and central Alabama, and then through northern Georgia, northern South Carolina, and eastern North Carolina. An increase in the probability of summer drought, with durations of up to three months, is also projected throughout the southern portion of the adjusted range. Yields are expected to be reduced over much of the range because of reduced soil water and changes in the edaphic substrates in the adjusted range. Regeneration practices probably will have to be modified to ameliorate the effects of the anticipated climatic change. Ripping, the use of containerized seedlings, and increased herbaceous weed control probably will be necessary under the expected changed conditions to assure satisfactory regeneration success. Tree-breeding programs should be dynamic, anticipating the possibility of range shift and reduced productive potential. Predictions of reduced productive potentials may be overly pessimistic in view of the possibility of increased water-use efficiency related to elevated carbon dioxide levels.

This paper addresses the changes in southern forest management practices that might be necessary as a result of rising atmospheric carbon dioxide (CO_2) and a probable climatic change. Basically, the general circulation model (GCM) data presented by Gibbs and Hoffman (this volume) and their climate change scenario under conditions of doubled atmospheric CO_2 were accepted as being within the realm of probability. It was assumed that changes would be gradual rather than catastrophic. Although most modelers and atmospheric scientists accept rising levels of CO_2 as a foregone conclusion, the magnitude and location of climatic change are still uncertain. However, based on the GCM data and the projections of

Solomon et al. (1984), climatic trends were examined on a regional and divisional basis and potential problems were identified. The effect of these problems on forest productivity and management activities was evaluated, and possible actions to alleviate or enhance forest productivity as a result of climatic changes were considered.

THE PHYSICAL PERSPECTIVE

The interaction of climatic change with the many possible ecosystem substrates of any area confounds the problem geometrically, and projected effects are, at best, only generalities. The general area of concern is the Southern Region, as defined by the U.S. Forest Service (1969), and more specifically, the area encompassed by the reported natural range of loblolly pine (*Pinus taeda* L.) (Figure 1). This broad area is characterized by diversity--diversity of physiography and land types and soils that have a profound influence both on the present range and productivity of loblolly pine and on the probable degree of impact of climatic change.

The majority of the soils supporting commercial production of loblolly pine are in the suborder Udults. The physical characteristics of the lands currently supporting loblolly pine vary widely with respect to soil properties, but they have several attributes in common:

- they are generally deep and well drained, with loamy surface horizons and clayey subsoil textures capable of supporting loblolly stands of good productivity; and
- they are, for the most part, easily accessible and able to withstand use of heavy equipment, making these landscapes physically well suited for intensive forest management.

THE BIOLOGICAL PERSPECTIVE

The effects that a changing climate might have on loblolly pine must first be examined by defining the annual growth cycle of the species and the climate variables that control this cycle. The second step is to examine how the projected changes in each of the "controlling" factors will affect each phenological phase.

A general outline of the phenological trends reported for terminal bud (height), diameter, and root growth of loblolly pine is illustrated in Figure 2. For established trees, terminal elongation does not occur during much of December, January, and early February, although some mitotic activity may be occurring at a slow rate (Cannell, 1978). The major environmental factor triggering the

Figure 1

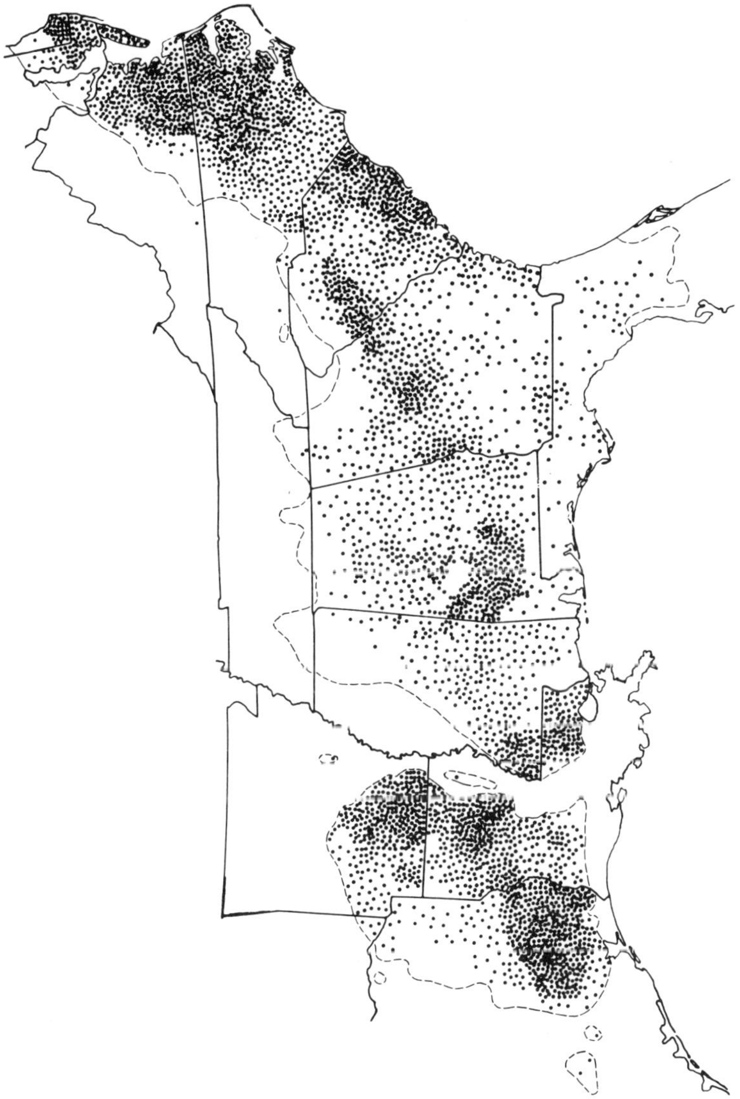

The natural range of loblolly pine (dashed lines) with commercial concentrations indicated by dots. Each dot represents 5 million cf (U.S. Forest Service, 1969).

Figure 2

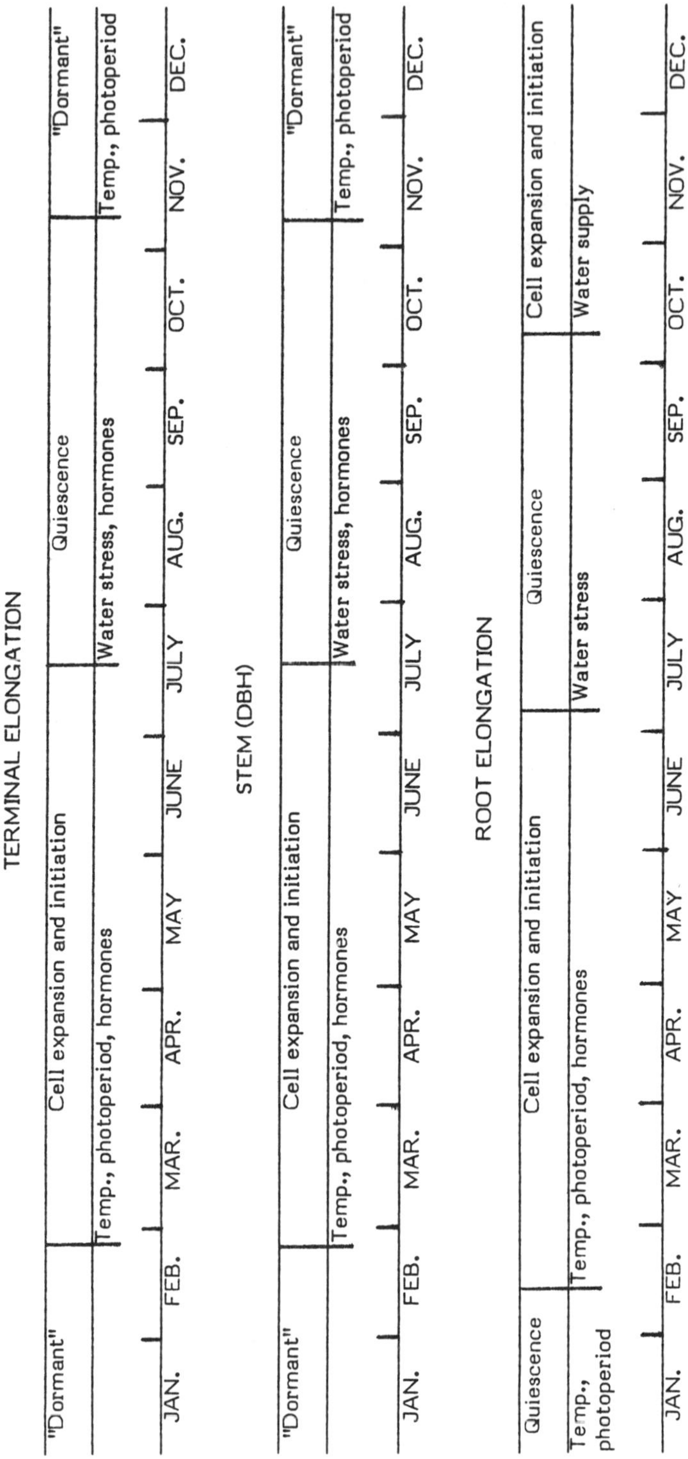

General phenological phases and controlling factors for three growth points of established loblolly pine trees (Cannell et al., 1976; Allen and McGregor, 1962).

cessation of this phase is temperature (Cannell at al., 1976). The threshold temperature, however, may differ between and within sources by 6°C (11°F) (Boyer, 1976). In the field, not much elongation of loblolly occurs until weekly temperatures average 6° to 10°C (43° to 50°F). The photoperiod is also important in controlling the shift from dormant to active growth mode (Allen and McGregor, 1962); this is especially true for northern sites (Allen and McGregor, 1962). Budbreak in the South usually occurs when the photoperiod approaches or exceeds 12 hours.

After budbreak occurs, both cell elongation and cell initiation begin. This phase begins in mid-February to early March and can continue into November, depending on geographical location and climatic conditions. In the early part of this phase, the rates of these two processes are limited by temperature (Ford, 1980).

The magnitude of realized height growth depends on how rainfall is distributed between the cool season (October to March) versus the warm (April to September) part of the growing season (Shoulders and Tiarks, 1980). If the average warm-season precipitation was less than 76 cm (29.9 inches), increased height growth was noted with increasing cool-season rainfall. However, if warm-season rainfall was greater than 76 cm, loblolly pine's height growth decreased with increasing cool-season precipitation. Loblolly heights increased with increasing warm-season rain until the sum of cool and warm (annual) rainfall exceeded 127 cm (50.0 inches). Slash pine (*P. elliottii* Engelm.) had a higher total annual rainfall tolerance (144 cm or 57 inches) than did loblolly before growth started to decrease with increases in summer precipitation. Shoulders and Tiarks' (1980) data set indicated that the optimum rainfall distribution for loblolly and slash pine's height growth was 61 cm (24 inches) with warm-season rainfall and 79 cm (31.1 inches) with cold-season rainfall.

In many areas within the loblolly range, terminal elongation is halted by internal water deficits (Zahner, 1962). Cell initiation is believed to be less sensitive to internal water deficits than cell expansion. However, if water deficits are of sufficient magnitude, cell initiation may be temporarily suspended (quiescence). If there is little stress during this phase, the phase will be terminated when the photoperiod falls in the range of 12 to 14 hours (Downs and Piringer, 1958) and temperature has decreased. The literature does not clarify the independent effect of photoperiod.

The cycle of diameter growth begins with the termination of dormancy in January and February. As with terminal elongation, it is questionable whether this is a true dormancy. However, Worrall (1980) suggests that there is a true cambial dormancy and that chilling and temperature interact to reinitiate cambial activity. In loblolly, the initial growth of earlywood is characterized by large cells and thin cell walls. The transition from earlywood to latewood (smaller cells and thicker cell walls) appears to be related to a decrease in upper crown photosynthate demand (cessation of height growth and needle expansion) and movement of photosynthate down the stem (Larson, 1969).

Inherent phenological patterns influence the percentage of earlywood versus latewood and, thus, specific gravity. Trees that break bud first should complete height growth and needle formation early in the growing season and make the transition to latewood production earlier in the growing season than trees that break bud late: thus, early budbreak trees would have a higher percentage of latewood than late flushing trees. This higher percentage of latewood would translate to woods of higher specific gravity and of higher quality for most purposes.

Bassett (1964) indicated that the period of latewood formation was controlled by moisture supply and could be manipulated through thinning. Moehring and Ralston (1967) indicated that both the level of moisture within the available range of a soil and the water-loss rates were important in determining growth of diameter, and, when soil water was generally about the midpoint of available soil water, diameter growth was only slightly affected by soil water content. However, when available water fell below this level, growth of diameter was greatly reduced. Moehring and Ralston further noted that the specific available water content at which growth was terminated varied with rate of moisture loss. For example, at a loss rate of 0.25 cm per day, growth of diameter ceased at an available water content equivalent to 4 bars of tension, but at a depletion rate of 0.64 cm per day, growth stopped at only 2 bars of the soil's water tension.

Several studies have indicated that the amount of diameter growth can also be closely correlated to the frequency of precipitation (Moehring and Ralston, 1967; Chang and Aguilar, 1980). Radial growth in the months of January through June is positively correlated with rainfall. This period includes the time of most rapid diameter growth. Langdon and Trousdell (1978) reported a strong correlation between annual basal area growth and cubic-foot (cf) volume growth with the departure from normal of rainfall during the April to October period. Positive departures resulted in higher than average yields. According to their equations, every 1 inch of rain above or below normal will add or subtract about 5 cf per acre per year in yield.

Manogaran (1973) integrated the effects or influence of rainfall, temperature, and soil moisture storage on yield of loblolly pine. The factors correlated with the monthly growth of volume were the actual evapotranspiration of the current month and the previous month, the actual evapotranspiration of the previous year for the period June through November, the soil's water deficit of the current month, and tree age. Manogaran and other researchers emphasize that not only is the current year's climate important for loblolly, but also antecedent climatic conditions must be considered to predict growth. This is probably because of the previous year's influence on stored photosynthate and, in conifers, the amount and condition of photosynthetic area carried over into the current year.

Little literature was found that dealt with trends of root growth for established loblolly pine trees. Kaufman (1977), however, reported trends for one- and two-year-old slash seedlings observed in the field and found that root growth starts before growth of either stem height or diameter. It is likely that roots do not have a dormant phase but instead are limited by low temperatures. Wenger's (1958) review indicates that optimum temperature for loblolly root growth is 25°C (77°F) and that root growth ceases when temperatures fall below 11°C (53°F) or exceed 30°C (86°F). Nambiar (1983) found that there was a ninefold difference between families in the number of new roots produced at 8°C (46°F) indicating that the ability to initiate new root apices is, in part, genetically controlled. Loblolly does initiate new roots in the temperature range of 7° to 9°C (44° to 48°F), but the rate is slow (Hennessey and Dougherty, 1984). Heninger and White's (1974) work with three conifer species indicated that the most optimum temperature for root growth was between 19° and 23°C (66° and 73°F).

CLIMATIC SCENARIOS

Data categorizing climate types by discrete divisions were obtained from ICF, Inc., and were calculated by the methodology of Gibbs and Hoffman (this volume). Divisions were selected to form an east-west and a north-south transect to represent temperature and precipitation patterns under conditions of doubled atmospheric CO_2 (year 2080).

Moving from southeastern Oklahoma and western Arkansas through east-central Mississippi and Georgia to coastal North Carolina, it was generally noted that the probability of drought during the growing season will markedly increase in the range's central portion. Summer precipitation is expected to decrease by

39% in western Arkansas, 17% in east-central Mississippi (Figure 3), and 4% percent in central Georgia; summer temperature rises of 5° to 8°C (9° to 14°F) are also anticipated. Fall precipitation will be more drastically reduced, ranging from 12% in Arkansas and 44% in Georgia to 61% in Mississippi. In general, winter and spring precipitation will increase, but because of the magnitude of the increase there will probably be an accompanying increase in runoff.

Coastal regions, however, may well fare better than the interior. Precipitation for coastal North Carolina is projected to decrease only about 1.2 inches (30 mm) from the historic average of 50.1 inches (1273 mm), a maximum decrease of 2.9 inches (51 mm), in July and 2.3 inches (58 mm) in September. In southwestern Louisiana, however, there will be an increase from 58.3 inches (1481 mm) to 65.3 inches (1659 mm), with most of the increase, 6.3 inches (160 mm), occurring during the warm season. Both warm-season and annual precipitation exceed the optimum for height growth of either loblolly or slash pine (Shoulders and Tiarks, 1980). Temperatures for coastal divisions are not expected to show the same rises as the interior, with temperatures during growing seasons increasing only 3° to 4°C (5° to 7°F) (Figure 4).

Moving northward into Pennsylvania and New Jersey, the climatic projections indicate that rainfall will remain relatively stable, or will show a slight increase, but that a major warming trend may occur. The major warming is expected in the winter with temperatures elevated 6° to 10°C (11° to 18°F). One additional trend analysis, of water balance, was made to assess loblolly distribution under the doubled-CO_2 climate scenario. For selected stations, Thornthwaite-Mather (1957) water balances were calculated and comparisons were made between the period 1951-1980 (normal), the "worst-drought" calendar year, and the year 2080 projections. The future for loblolly looks grim indeed on shallow or sandy soils in eastern Oklahoma and western Arkansas. A 4-inch storage capacity was selected to represent these critical sites. The 30-year average (1951 to 1980) for southeastern Oklahoma indicates that a normal water deficit (potential versus actual evapotranspiration) of about 5.5 inches (140 mm) will occur in June, July, and August, with an annual precipitation of 46.1 inches (1172 mm) (Figure 5). The worst-case year for seedling survival in the recent past was 1980 (Dougherty, 1981). With the same soil storage capacity, a deficit of 11.6 inches (295 mm) occurred from June through August, with precipitation of 44.6 inches (133 mm) for that year (Figure 6). A storage capacity of 10 inches (254 mm), which would correspond to the deeper soils in the area,

Figure 3

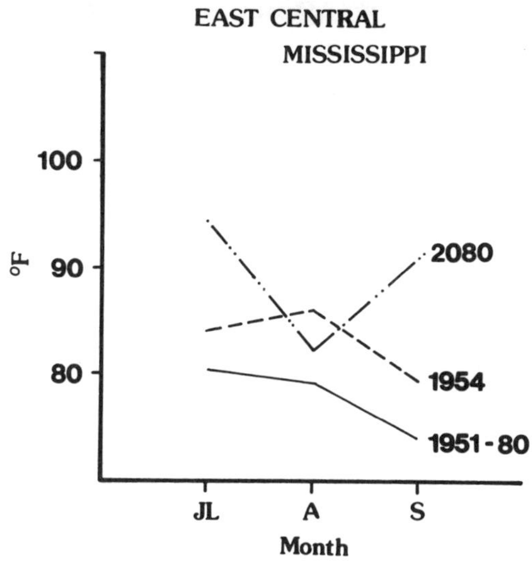

Average monthly temperatures for July, August, and September for east-central Mississippi: 30-year normal, 1954, and 2080 projection.

Figure 4

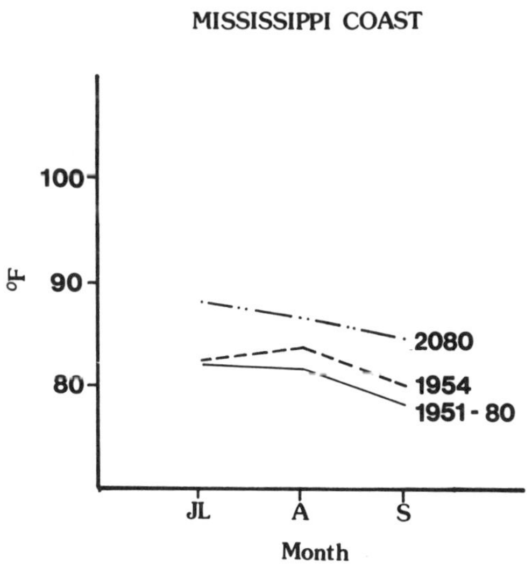

Average monthly temperatures for July, August, and September for coastal Mississippi: 30-year normal, 1954, and 2080 projection.

Figure 5

OKLAHOMA
1951 - 1980

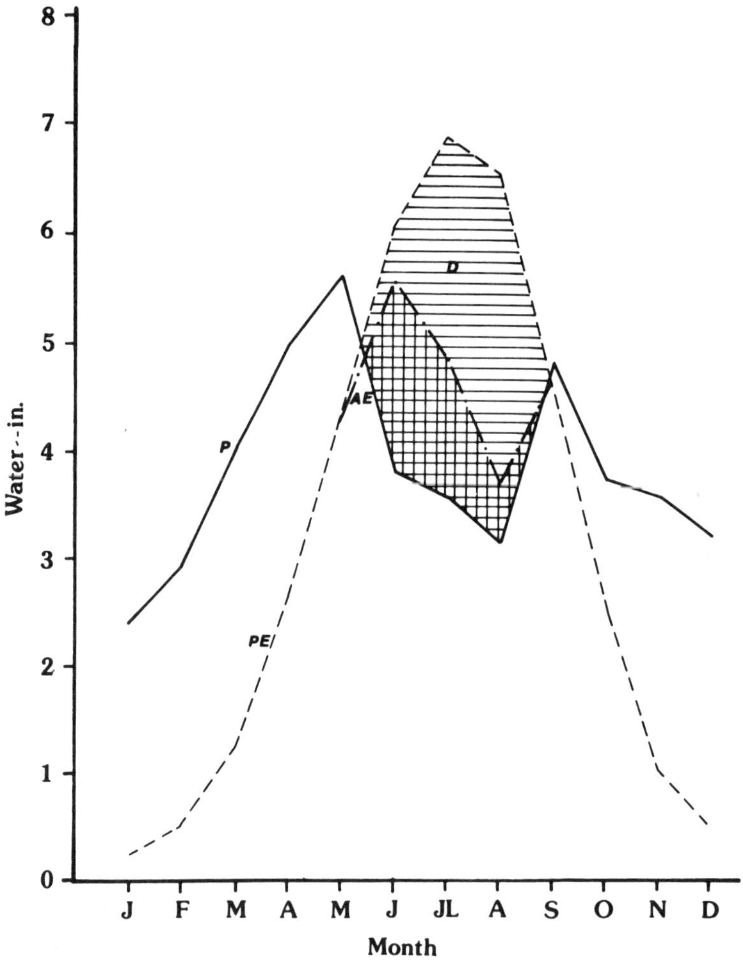

Water balance for southeast Oklahoma for the period 1951 to 1980, assuming a growing-season deficit of 5.5 inches with a 4-inch storage and annual precipitation of 46.1 inches. P = precipitation, PE = potential evapotranspiration, AE = actual evapotranspiration, D = water deficit, cross-hatched area = removal from soil storage.

Figure 6

BROKEN BOW
1980
4″ storage capacity

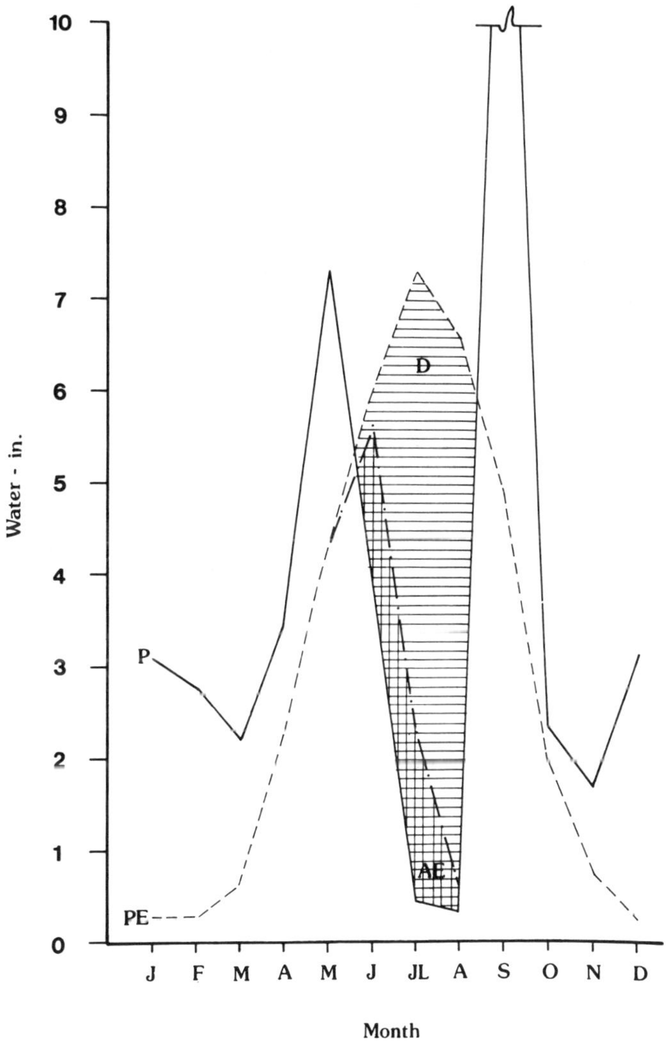

Water balance for worst-case drought year, 1980, in Broken Bow Dam, Oklahoma, assuming a growing-season deficit of 11.6 inches with a 4-inch storage and annual precipitation of 44.6 inches.

resulted in a deficit of 7.5 inches (190 mm) for 1980. By contrast, the year 2080 scenario would result in a deficit of 20.6 inches (523 mm) between June and October, with total precipitation of 43.1 inches (1095 mm) (Figure 7).

East-central Mississippi will also undergo a dramatic increase in water stress. The normal rates for 1951 to 1980 are 55.3 inches (1405 mm) of precipitation, with an average deficit of 4.2 inches (107 mm) and a 4-inch storage (102 mm) from June through September (Figure 8). The 2080 scenario indicates that, although annual precipitation may remain essentially the same (56.9 inches or 1445 mm), the distribution (by month) will differ considerably (Figure 9), and the deficit will reach 24 inches (610 mm) from May through October. Even deep silty soils with a 12-inch (305 mm) storage capacity will have a deficit of 17.3 inches (439 mm). By contrast, the worst-case scenario for the past 30 years was in 1954, with total precipitation of only 39.4 inches (1001 mm) and a deficit of 16.7 inches (424 mm) for a 4-inch storage capacity of soil (Figure 10).

POTENTIAL IMPACTS OF CLIMATIC CHANGE

The likely effects of projected climatic alterations on loblolly range, productivity, and biological integrity were deduced, and the impact of these findings on established stands is discussed.

Changes in Range and Productivity

Based on the work of Solomon et al. (1984), the range of loblolly pine will undergo an appreciable shift to the north and northeast, with a smaller withdrawal from west to east (Figure 11). The present range of loblolly encompasses about 344,000 square miles (139,271 hectares), of which 233,000 square miles (94,332 hectares) are considered to be commercial loblolly pine lands (Sheffield and Knight, 1982; McWilliams and Birdsey, 1984). The range projected by Solomon et al. (1984) for the year 2080 shifts and expands to a total area of about 411,000 square miles (166,397 hectares).

The addition of the water-balance calculations to the projections of Solomon et al. (1984) indicates that the western extent of the range might well withdraw into central Arkansas and extend northward along the Upper Piedmont and into eastern Pennsylvania and New Jersey. This adjusted range would equal about 385,000 square miles (155,870 hectares). Unfortunately, it is unlikely that the same proportion of commercial loblolly lands will be retained, because an appreciable proportion of the "new" range will be in

Figure 7

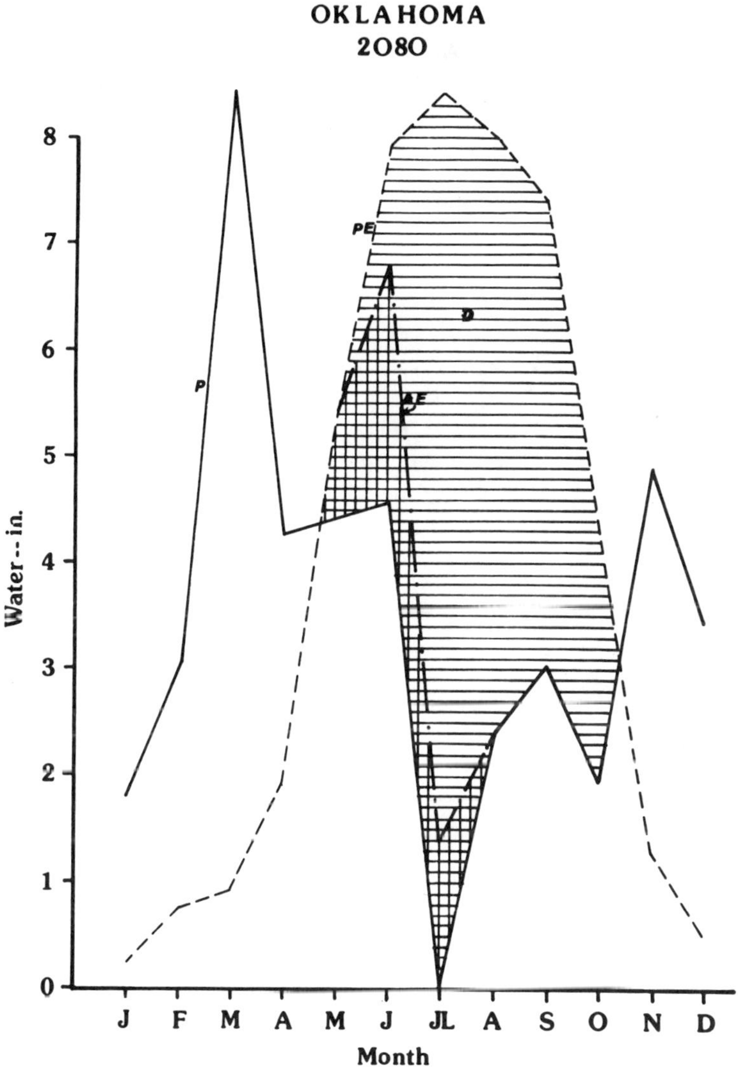

Water balance for the year 2080 in southeast Oklahoma, assuming a growing-season deficit of 20.6 inches with a 4-inch storage and annual precipitation of 43.1 inches.

Figure 8

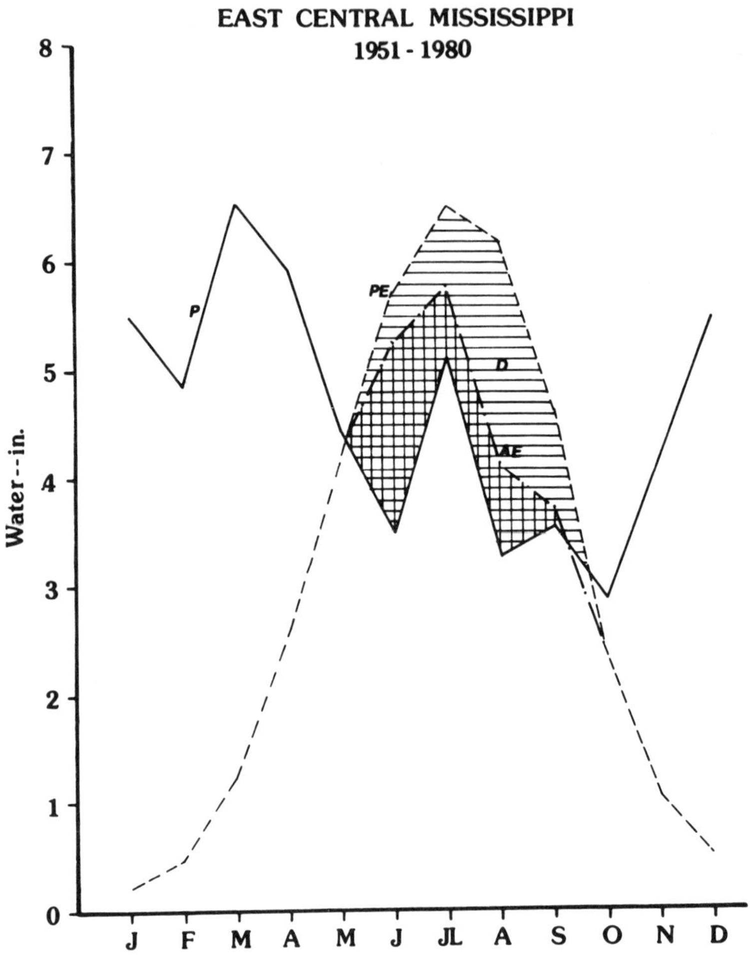

Water balance for the east-central Mississippi 30-year normal, assuming a growing-season deficit of 4.2 inches with a 4-inch storage and annual precipitation of 55.3 inches.

Figure 9

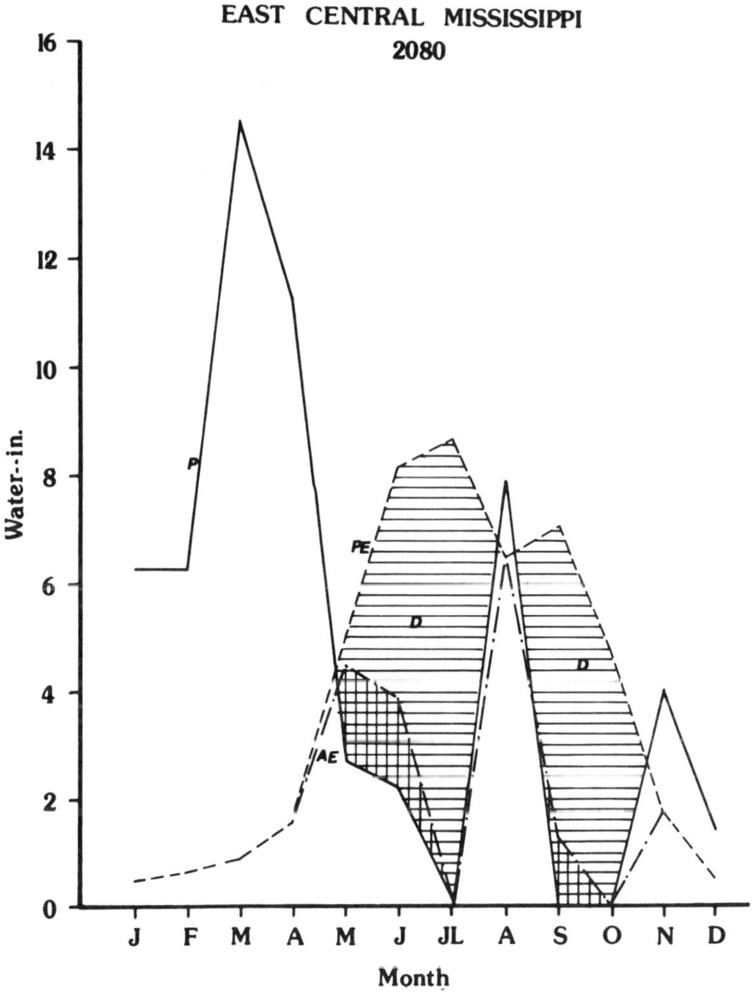

Water balance for east-central Mississippi, 2080 projection, assuming a growing-season deficit of 24 inches with a 4-inch storage and annual precipitation of 56.9 inches.

Figure 10

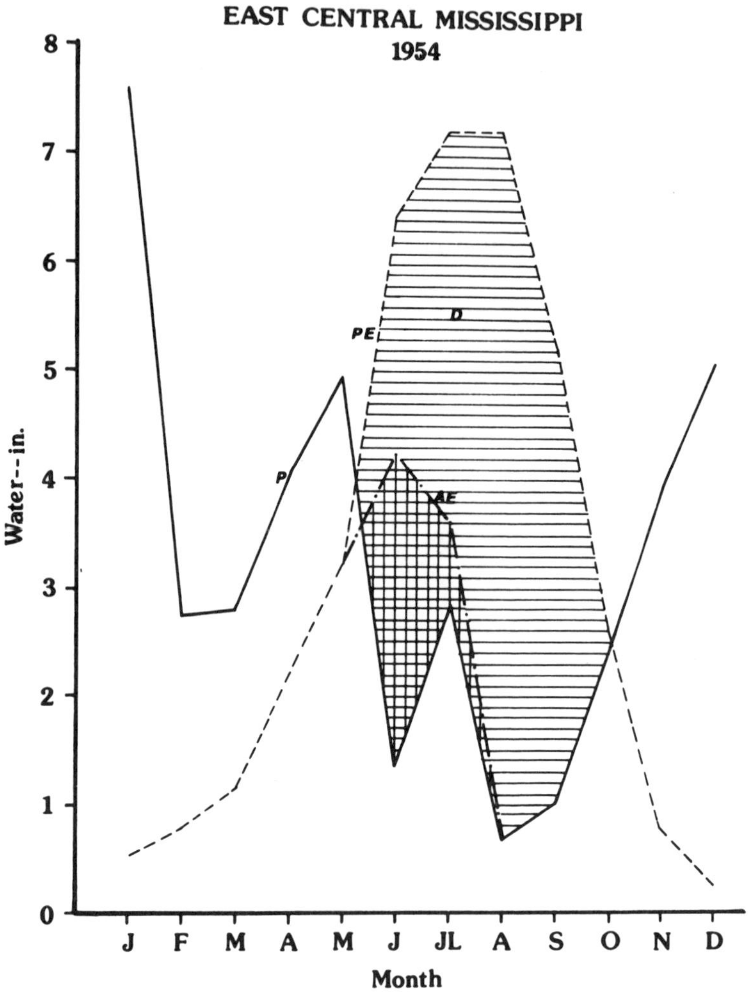

Water balance for east-central Mississippi, 1954 drought, assuming a growing-season deficit of 16.7 inches with a 4-inch storage and annual precipitation of 39.4 inches.

Figure 11

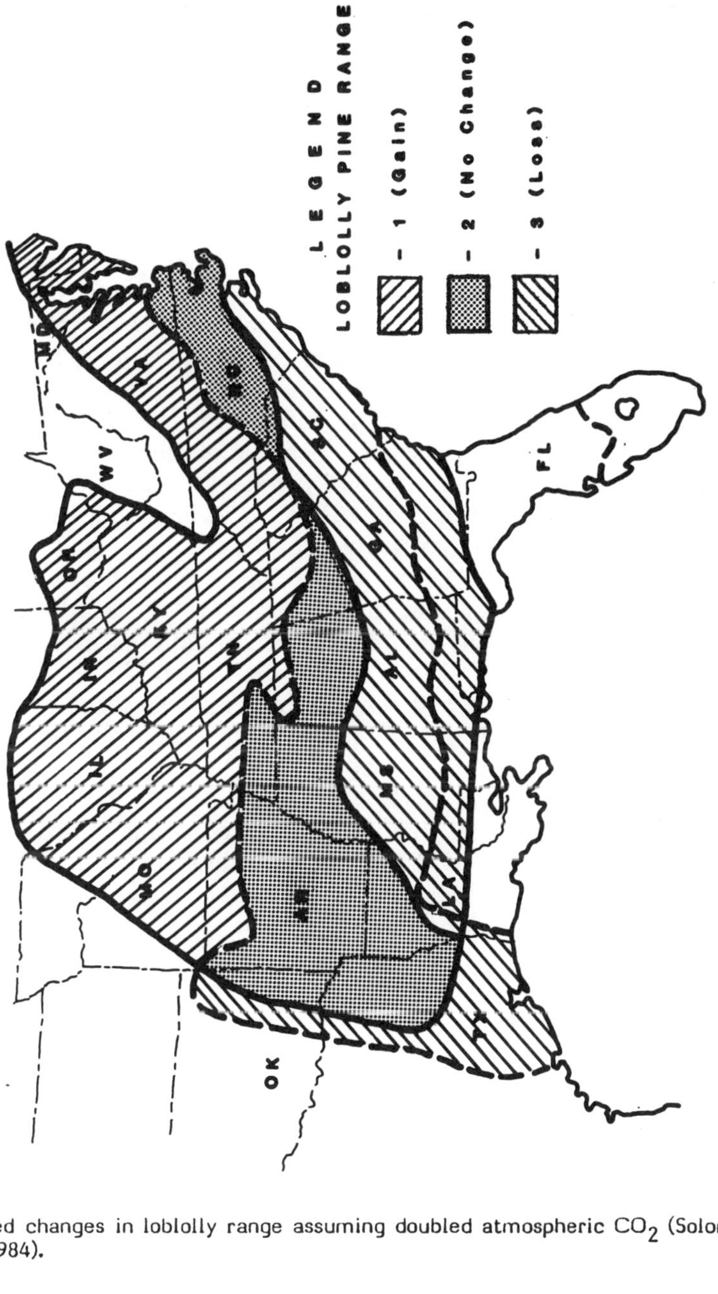

Projected changes in loblolly range assuming doubled atmospheric CO_2 (Solomon et al., 1984).

the interior highlands and plateaus.

This potential movement has profound implications for future forest management; no longer will the forest manager be able to enjoy the relatively easy access to loblolly timber that is now enjoyed in the coastal plain. Nor will many of the current logging and regeneration practices be directly transferable to the highlands. In contrast to the coastal plain, the highlands are deeply dissected, with steep slopes and shallow soils. And if, in fact, the range extends into the central lowlands, it is decidedly unlikely that land use would change from crop to forest.

Other implications are perhaps not so obvious, particularly with respect to changes induced in soils. As stated earlier, the bulk of commercial pinelands are comprised of relatively deep soils, whereas areas into which loblolly is projected to move have a relatively high proportion of shallow, steeply sloping, coarse-textured soils (Smith et al., 1973) of the steep and deeply dissected rocky uplands. These soils and landscapes, because of their poorer regimes, may support only marginal loblolly pine at best. Accordingly, a general decline in overall productivity is anticipated, even though the area of the range has increased. In the mid-South, loblolly currently occupies the better sites: that is, about two-thirds of the loblolly land with a site index (base age, 50 years) of 80 feet (24 m) or more. Loblolly plantations occupy approximately 64% of the loblolly acreage in the interior highlands. These plantations have generally been planted on the better sites of the area, more so than in the coastal plains, and yet the reported average site index (50 years) is only 75 feet (23 m) (McWilliams and Birdsey, 1984).

Other soil problems likely to emerge as the climatic trend continues are: (1) increased acidity; (2) increased aluminum toxicity; (3) increased degradation of soils of the middle and lower coastal plain due to formation of plinthite "pans"; (4) reduction in soil organic matter; and (5) increased area of problem soils with high salt (Na) content (Pettry, personal communication). In general, those soils that now contain an appreciable amount of weatherable minerals will probably be the most drastically affected by the anticipated climatic change because such soils have a greater potential for weathering impacts. Loessial soils in Louisiana, Mississippi, Arkansas, and Tennessee, for example, may well become more acid. With leaching of calcium, the soil structure would be degraded, and with increased winter/spring precipitation, massive slumps and land movement could occur.

What will happen in the present southern extent of the loblolly range? Will it be replaced by slash pine? Will the coastal areas and the southern portion of the middle coastal plain move toward savanna conditions caused by elevated temperatures and precipitation and an increase in plinthite development? The more likely scenario is for loblolly to be confined to either excessively, somewhat excessively, or well-drained soils in the middle coastal plain. However, at present there is not enough information to hazard even a guess on the future direction in this area.

Another factor that must be considered in future management activities is the probability of greatly increased erosion. With most of the southern region projected to have increases in spring precipitation, the scheduling of field operations may have to be modified. Projected increases in March and April rainfall in east-central Mississippi, for example, are about 228 and 215% respectively. As climate becomes more limited for growth, erosional losses of soil become even more critical.

In summary, the lands presently producing the bulk of the wood fiber may be lost to future loblolly production or, at best, become only marginal. Moreover, the lands onto which loblolly is expected to move are generally of lower quality as sites for loblolly and present greatly increased management problems and hazards.

Biological Effects

Based on the anticipated climatic changes, one of the most obvious problem areas will be the regeneration phase of management. There are, however, other biological aspects that require consideration. Because the base data for climatic change are tenuous, potential problem areas are only briefly overviewed.

Seedlings

Although seedling morphology can be modified through nursery practices, the question still remains as to whether any form of bare-root planting stock will survive the adverse summer/fall environmental conditions during the first growing season in the field. There are, however, a few offsetting factors, such as growth-positive increases in winter and spring temperatures. Although higher summer and early fall temperatures will generally decrease survival and growth, it is possible that photoperiod-caused growth cessation at 12 hours of daylight in the autumn will be delayed by elevated temperatures (Downs, 1962). Notwithstanding the positive influences of predicted increased atmospheric CO_2 and climate change on survival and growth, the overriding consideration will

Figure 12

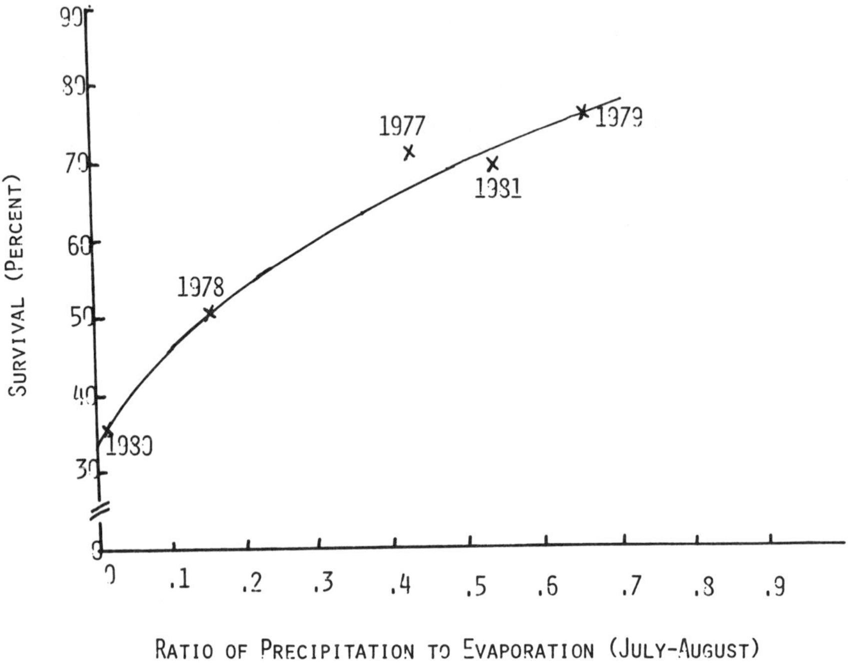

The relationship of average percentage of seedling survival and the ratio of precipitation to evaporation for July/August in the western part of the loblolly range (Barnett et al., 1984).

be extended periods of water and temperature stress during the first field season.

The effect on seedling survival that the predicted shift in climate will have is inferred from a plot of expected survival against an index of moisture supply and evaporative demand for the critical establishment months of July and August (Figure 12). Many areas experiencing 80% to 90% survival with current bare-root techniques will drop to a survival rate of 35 to 50% unless additional steps are taken to plant trees selected and developed to survive in droughty environments. Much can be done to precondition loblolly seedlings to survive in drought-ridden environments. Also, the range of genetic traits that confer survival under droughty conditions, such as stomatal characteristics and behavior, has not been fully explored. Of major concern in seedling survival are the high temperatures predicted to occur in the months of July, August, and September. No mortality rates for loblolly seedlings under this type of temperature-moisture regime are reported.

The worst year for seedling survival across most of the western loblolly range was 1980. The duration and intensity of drought experienced in 1980 was no worse than that experienced in previous drought years, but the unique feature of this year, at least in southeastern Oklahoma and east-central Mississippi, was the number of days on which the maximum temperatures reached or exceeded 100°F (38°C): 24 and 19 days, respectively, during July and August, plus an additional 17 days, had a maximum temperature of 98° or 99°F (approximately 37°C). The average temperature for July and August 1980 was 83.3°F (28°C), about 3.2°F (1.8°C) above average. The predicted average for July and August 2080 is 91.2°F (33°C). The duration of high temperatures predicted for 2080 is also much longer than that in 1980. For east-central Mississippi, the average temperature in July and August 1980 was 85.3°F (29.6°C), with a positive departure of 4.8°F (2.7°C) from the 30-year normal. The projected temperature for July and August 2080 is 88.3°F (31.3°C).

Undoubtedly, seedling survival will be a major concern under these conditions. It is unlikely that elevated CO_2 levels will be of much benefit under extended periods of low soil moisture and high temperatures because stomatal resistance will be extremely high most of the time. The effects of extended periods of high temperature on seedlings under water stress need better quantification.

The use of cultural treatments that conserve moisture will become more important in establishing stands able to withstand the decrease in rainfall and the increase in temperature. These treatments include subsoiling, control of herbaceous competition, and brush control. Containerized seedlings may play an important role by reducing the transplant shock and thus permitting more rapid growth early in the season. This would insure that the seedlings were better established before the hot, dry period began.

One positive factor with respect to either containerized or bare-root seedlings is the anticipated increase in water-use efficiency with increasing CO_2 levels. Pearcy and Bjorkman (1983) indicate that, under greenhouse conditions, C_3 plants nearly doubled their water-use efficiency in a CO_2-enriched (660 parts per million) atmosphere.

What then are the options? Is technology currently available to overcome the environmental stresses? If the low survival cannot be offset, even on deep soils, then a worst-case migration of forest industry from eastern Texas, southern and central Louisiana, Mississippi, Alabama, Georgia, and Florida might be expected if the projections of Solomon et al. (1984) are correct.

Established Stand Effects

A shift in climate will affect meristematic activity of the shoot and root systems in different ways. The phenology of these systems and the environmental factors reported to control each phase are illustrated in Figure 2. The effect on existing stands of a shift toward warmer, drier summers and moister springs and falls is discussed below.

Root Growth

Elevated temperature in the late fall-winter period would be expected to improve root growth during these periods. Low temperature is the factor that normally limits root growth during this period. However, in poorly drained soils, reduced aeration may be the limiting factor instead of low temperatures. Higher rainfall amounts in the winter-spring period would reduce root growth on these sites. Greater additions of moisture in this period will cause a shift from loblolly pine to more flood-tolerant species on sites currently having marginal drainage for loblolly.

In the summer-early fall period, moisture stress would be expected to limit root growth throughout much of the region, although no threshold value could be located in the literature. Wenger's report (1958) suggests that high temperatures (greater than 30°C or 86°F) could also inhibit loblolly root development. If so, root growth in the summer period would be depressed considerably under the 2080 climate scenario.

Apical Growth

Height growth, illustrated in Figure 2, is initiated after dormancy when temperatures begin to average above 6°C (43°F). With a shift to warmer spring temperatures, height growth would be expected to start earlier in the season and to progress faster. Lack of chilling temperature may delay budbreak for families that require chilling. The first flush consists largely of expansion of terminal bud cells formed the previous fall-winter. Thus, the conditions that existed in the previous late summer and fall dictate, to some extent, the amount of growth expected from the first flush. Loblolly pine has a recurrent flushing pattern of shoot growth and would be expected to continue to grow until limited by internal water stress.

No lower thresholds of available water content in soils or of internal water potential to stop height growth have been reported. Moehring and Ralston (1967) and Bassett (1964) reported that basal area growth ceased when the available soil water was depleted to approximately 40% of the maximum. This threshold value for height growth cessation predicts stopped growth by early June across much of the loblolly range. With the high evaporative demands and low precipitation inputs predicted for July, August, and September, the number of flushes would likely be reduced across much of the range of loblolly pine.

The effects of higher rainfall amounts in the spring could be negative if solar radiation is below the level needed for high rates of photosynthesis to occur. This would be critical if previous summer-fall conditions were not favorable for building a reserve of stored photosynthates.

Based on the projected 2080 climate scenario of doubled atmospheric CO_2, conditions for cell expansion and cell initiation in the early season would diminish for height growth. It is more likely that previous-year conditions, which influence the number and development of terminal bud cells available for expansion in the first flush and the amount of stored reserves available for synthesis of new growth, will reduce observed height growth throughout much of the loblolly range. As previously discussed, the number of flushes will probably decrease, decreasing productivity.

Cambial Growth

Cambial meristematic tissue begins its growth cycle after dormancy just as height does (Figure 2). Dormancy is terminated as photoperiod and temperature increase. As spring temperatures rise, growth of diameter can be expected to begin earlier in the year as long as photoperiod and chilling requirements are met. With a 6°C (11°F) threshold for growth initiation, growth in diameter would start from four to eight weeks earlier in the year than it does now.

Diameter growth can occur into the autumn when it is terminated by short days or low temperatures. However, most areas experience a period of quiescence caused by internal water stress, which halts cell expansion and perhaps cell initiation. In the northwest part of the range, growth of the basal area can be halted by drought by early June. Only approximately 50% of the basal area growth will be attained under these conditions versus what can occur in years in which rainfall is not limiting (Bassett, 1964). Based on the 2080 climate scenario for Oklahoma, growth cessation of the basal area is predicted in June.

Not only would diameter growth be affected, but also a shift in quality of wood may result. The shift from production of earlywood to latewood occurs when most shoot and needle expansions are near completion. If shoot and needle growths start earlier, then the shift would occur earlier. However, the shortening of the period for latewood production would mean a higher percentage of earlywood than latewood. The result is a harvest of lower-specific-gravity wood and, thus, wood of poorer quality. This effect would be greater on loblolly pine in areas expected to experience the more extended periods of drought than in the coastal part of the range. This impact could be reduced or reversed if, indeed, elevated atmospheric CO_2 levels substantially increased the amount of photosynthate produced during the early season. The amount of photosynthate available for downward translocation is believed to be a key factor in the shift from earlywood formation to latewood formation.

Across much of the loblolly range, growth in height and diameter is likely to decrease. Because these are the major determinants of volume, volume too would decrease. Manogaran (1973) suggested that a decrease of approximately 5 cf per acre per year can be expected for each 1 inch of rainfall below normal experienced for the period of April through October. By this rule, volume would change from +30 cf to -35 cf per acre per year from the southwest coast to the northwestern part of the loblolly range. From the drought standpoint, this underestimates volume reduction because evaporative demand would be much higher than that of the period from which Manogaran's original data were collected.

Elevated CO_2 levels should increase loblolly productivity early in the season. Thus, in range areas where early season conditions become more favorable and little change in summer temperature and rainfall occurs, a net gain in productivity is expected. How then are the growth rates that might be expected under the future climatic scenario predicted?

The detailed climatic data necessary for estimates of growth from soil-water relationships are not available, and it is not thought that valid growth projections can be made at this time. However, net annual above-ground productivity (NAAP) can be calculated by means of its relation to annual actual evapotranspiration (AE) (Rosenzweig, 1968). Basically, all this calculation can accomplish is a relative measure of differences or trends. The relationship, $\log_{10}$ NAAP (grams per square meter) = (1.66 + 0.27) $\log_{10}$ AE (mm) - (1.66 - 0.07), assumes a constant CO_2 concentration of 0.03% and thus somewhat underestimates the year 2080 productivity for any station.

The results (Figure 13) indicate that southeastern Oklahoma loblolly will decrease in productivity by about 30% but east-central Mississippi's stand will undergo a 47% reduction. However, with a mean site index (50 years) for loblolly pine of 84 feet (24.0 m) in the coastal plain and an average of 75 feet (31.5 m) in the highlands, the two areas should have about equal yields of wood fiber. Also of interest is the projected productivity of eastern Pennsylvania in the year 2080--1.5 tons per hectare higher than the present normal for east-central Mississippi, and about equal to coastal Mississippi during the drought of 1954. The productivity of central Tennessee will fall between that of east-central Mississippi and southeast Oklahoma--all quite low compared with present productivity.

It should be noted that in the case of Oklahoma and interior Mississippi, the relationship does not allow for the very real possibility that high temperatures alone during July, August, and September may be the most limiting growth and survival factor. If so, then the relationship illustrated in Figure 13 may be useful only for estimating the productivity in the northward extension of the range, that is, Tennessee and Pennsylvania. The relationship also does not account for tree-breeding programs that can select for drought resistance in a relatively short time, thus maintaining or increasing productivity on droughty sites.

In sum, available information predicts an appreciable decrease in both seedling survival and productivity in the southern and western portions of the range of loblolly pine in the year 2080. Application of existing technology, such as containerized seedlings, herbaceous weed control, and site preparation methods to increase rooting depth and to retain topsoil, will probably be required

Figure 13

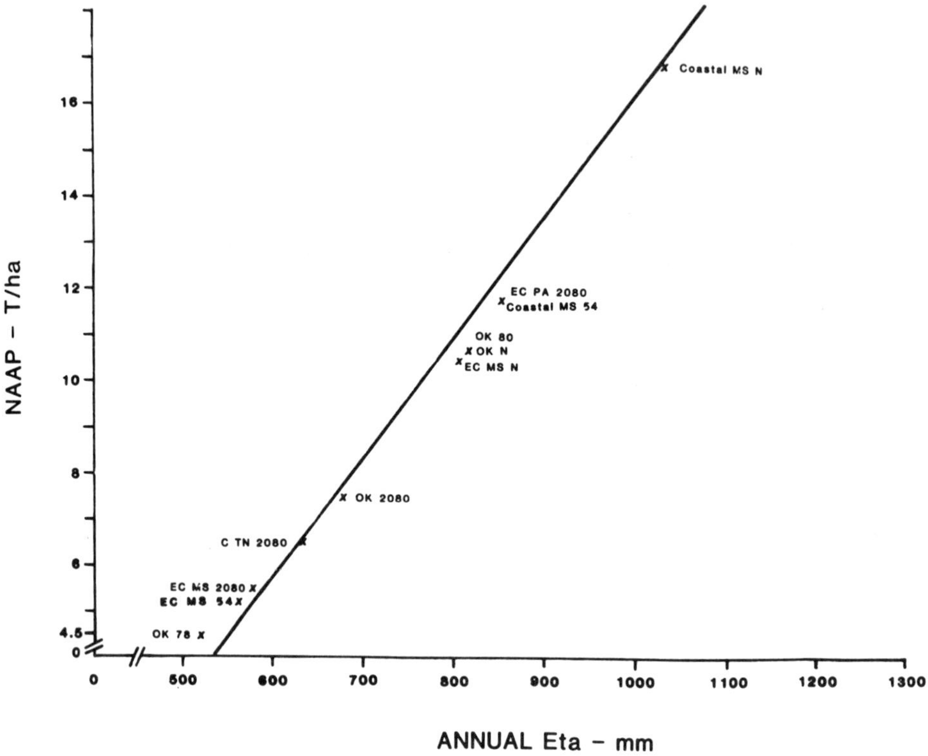

The relation of net above-ground productivity (tons per hectare) to annual actual evapotransipiration (in millimeters) for selected stations/divisions and years (Rosenzweig, 1968).

throughout the range to insure survival. Productivity levels will generally decrease because more upland sites will become marginal in terms of net annual productivity; a water-holding capacity of 4 to 6 inches (102 to 152 mm) probably will not support commercial production of loblolly pine under the new climatic regime.

These projections force consideration of carbon dioxide enrichment (CO_2 fertilization) and of possibly decreased rotation length, at least on deeper soils, of intensively managed stands (Strain and Bazzaz, 1983). A survival-positive increase in photosynthetic water-use efficiency resulting from decreasing stomatal conductance as CO_2 increases can also be anticipated.

CONCLUSIONS

There is little doubt that atmospheric CO_2 levels are increasing. What is in doubt is the interrelationship between rising CO_2 and climatic change with respect to variations in both temperatures and precipitation amounts and the geographical location and extent of the fluctuations. The rate of change is equally uncertain: Will the increase be linear or exponential? Will increased water efficiency offset anticipated drought and temperature stresses? If the answers to these questions were known, then future range limits and productivity estimates might be projected with some confidence. However, only the possible effects of one climatic scenario, with all of the uncertainty this implies, have been projected. But from this scenario it is evident that even if the water deficits are only 70% or even 50% as bad as indicated, problems in regenerating and maintaining productivity of loblolly pine will still be present.

It is obvious from the response to this conference that this research area is not of pressing importance to the forest industry for a number of reasons, two of which are short-range planning horizons and the belief that managers are sufficiently flexible to change management practices when climatic change occurs. However, this posture necessarily assumes a linear change, not a more catastrophic, exponential one. In fact, there are research areas that will not only benefit current management practices and increase growth and yield, but will also provide the basis or flexibility to increase or modify future research to meet changing climate.

Areas of research that deserve more attention are:

- increased knowledge of the reaction of loblolly pine and its competitors to elevated atmospheric carbon dioxide, particularly with respect to growth initiation, maintenance, and cessation;

- development of new, low-cost, and environmentally safe herbicides for both woody and herbaceous species;
- additional research efforts on regeneration techniques for droughty sites--including development or modification of equipment for harvesting and site preparation on steep slopes and shallow soils and development of containerized seedling handling; and
- an increase in the gene pool and use of intensive genetic selection for drought resistance and early flushing.

ACKNOWLEDGMENT

We thank John L. Tingle, research associate, Mississippi Remote Sensing Center, for calculating the water-balance data and providing editorial comment.

REFERENCES

Allen, R. M., and W. H. D. McGregor. 1962. Seedling growth of three southern pine species under long and short days. *Silv. Genet.* 11, 43-45.

Barnett, J. P., T. E. Campbell, and P. M. Dougherty. 1984. Seedling establishment-artificial methods. *In Proceedings of a Symposium on the Loblolly Pine Ecosystem (West Region).* Jackson, Miss.

Bassett, J. R. 1964. Tree growth as affected by soil moisture availability. *Soil Sci. Soc. Am. Proc.* 28, 436-438.

Boyer, W. D. 1976. Thermal efficiency: a possible determinant of height growth potential in young loblolly pines. *Forest Sci.* 22, 279-282.

Cannell, M. G. R. 1978. Genetic variation in seasonal growth phenology within *Pinus taeda* L.: literature review and seedling studies on date of bed set. *Weyerhaeuser Tech. Rep.*

Cannell, M. G. R., S. Thompson, and R. Lines. 1976. An analysis of inherent differences in shoot growth within some north (temperature) conifers, p. 173-205. *In* M. G. R. Cannell and F. T. Last (ed.), *Tree Physiology and Yield Improvement.* Academic Press, Inc., New York.

Chang, M., and G. J. R. Aguilar. 1980. Effects of climate and soil on the radial growth of loblolly pine (*Pinus taeda* L.) in a humid environment of southeastern U.S.A. *Forest Ecol. Mgt.* 3, 141-150.

Dougherty, P. M. 1981. Unpublished data. Wright City Field Research Station, Weyerhaeuser Co., Wright City, Okla.

Dougherty, P. M. 1984. Unpublished data. Wright City Field Research Station, Weyerhaeuser Co., Wright City, Okla.

Downs, R. J., 1962. Photocontrol of growth and dormancy in woody plants, p. 133-148. *In* T. T. Kozlowski (ed.), *Tree Growth.* Ronald Press Co., New York.

Downs, R. J. and A. A. Piringer. 1958. Effects of photoperiod and kind of supplemental light on vegetative growths of pines. *Forest Sci.* 4, 185-195.

Ford, E. D. 1980. Impact of environment on the shoot elongation of conifers: short term effects, p. 107-126. *In Control of Shoot Growth in Trees.* Proceedings of the Joint Workshop of IUFRO Working Parties on Xylem and Shoot Growth Physiology. Fredericton, New Brunswick, Canada, July 1980.

Heninger, R. L., and D. P. White. 1974. Tree seedling growth at different soil temperatures. *Forest Sci.* 20, 363-367.

Hennessey, T. C., and P. M. Dougherty. 1984. Characterization of the internal water relations of loblolly pine seedlings in response to nursery cultural treatments: implications for reforestation success, p. 225-244. *In* M. L. Durea and G. N. Brown (ed.), *Seedling Physiology and Reforestation Success.* Martinus Nijhoff/D.W. Junk Publishers, The Hague.

Hodgkins, E. J., M. S. Golden, and W. F. Miller. 1979. Forest habitat regions and types on a photomorphic-physiographic basis. So. Coop. Series 210. Alabama Agricultural Experiment Station, Auburn, Ala.

Kaufman, C. M. 1977. Growth of young slash pine. *Forest Sci.* 23, 217-226.

Langdon, O.G., and K.B. Trousdell. 1978. Stand manipulation: effects on soil moisture and tree growth in southern pine and pine hardwood studies, p. 221-236. *In* W. E. Balmer (ed.), *Proceedings Soil Moisture ... Site Productivity Symposium.* Myrtle Beach, S.C., November 1-3, 1977. U.S.D.A. Forest Service.

Larson, P. R. 1969. Wood formation and the concept of wood quality. Bull. no. 74. Yale University School of Forestry, New Haven.

Manogaran, C. 1973. Economic feasibility of irrigating southern pines. *Water Resources Res.* 9(6), 1485-1496.

McWilliams, W. H., and R. A. Birdsey. 1984. Loblolly pine resources of the Midsouth. *In Proceedings of a Symposium on the Loblolly Pine Ecosystem (West Region)*, Jackson, Miss. (in press).

Moehring, D. M., and C. W. Ralston. 1967. Diameter growth of loblolly pine related to available soil moisture and rate of soil moisture loss. *Soil Sci. Soc. Am. Proc.* 31, 560-562.

Nambiar, E. K. S. 1983. Root development and configuration in intensively managed *Radiate* pine plantations. *Plant Soil* 7, 37-47.

Pearcy, R. W., and O. Bjorkman. 1983. Physiological effects, p. 83, 84. *In* E. R. Lemon (ed.), *CO_2 and Plants--The Response of Plants to Rising Levels of Atmospheric Carbon Dioxide.* Westview Press, Inc., Boulder, Colo.

Perkins, H. F., H. J. Byrd, and F. T. Richie, Jr. 1973. Utisols--light colored soils of the Warm Temperate forest lands, p. 73-77. *In* S. W. Buol (ed.), *Soils of the Southern States and Puerto Rico.* So. Coop. Service 174. N.C. Agricultural Experiment Station, Raleigh, N.C.

Rosenzweig, M. L. 1968. Net primary productivity of terrestrial communities: prediction from climatological data. *Amer. Nat.* 102, 67-74.

Sheffield, R. M. and H. A. Knight. 1982. Loblolly pine resource-southeast region, p. 7-24. *In Proceedings of a Symposium on the Loblolly Pine Ecosystem (East Region)*, Raleigh, N.C.

Shoulders, E., and A. E. Tiarks. 1980. Predicting height and relative performance of major southern pines from rainfall, slope and available soil moisture. *Forest Sci.* 26(3), 437-447.

Smith, R. M., D. E. Pettry, and C. Breinig. 1973. Inceptisols--weakly developed soils, p. 33-42. *In* S. W. Buol (ed.), *Soils of the Southern States and Puerto Rico.* So. Coop. Series 174. N.C. Agricultural Experiment Station, Raleigh, N.C.

Solomon, A. M., M. L. Tharp, D. C. West, G. E. Taylor, J. W. Webb, and J. L. Trimble. 1984. Response of unmanaged forests to CO_2-induced climate change: available information, initial tests, and data requirements. DOE/NBB/0053 TR009. U.S. Department of Energy, Office of Energy Resources, Washington, D.C.

Strain, B. R., and F. A. Bazzaz. 1983. Terrestrial plant communities, p. 179, 193. *In* E. R. Lemon (ed.), *CO_2 and Plants--The Response of Plants to Rising Levels of Atmospheric Carbon Dioxide.* Westview Press, Inc., Boulder, Colo.

Thornthwaite, C. W., and J. R. Mather. 1957. Instructions and tables for computing potential evapotranspiration and the water balance. Drexel Institute of Technology Laboratory of Climatology, Publications in Climatology, vol. X, 3. Centerton, N. J.

U.S. Forest Service. 1969. A forest atlas of the south. Southern and Southeastern Forestry Experiment Stations, New Orleans, La. and Asheville, N.C.

Wenger, K.F. 1958. Silvical characteristics of loblolly pine. Paper no. 98. U.S.D.A. Forest Service. Southeast Forestry Experiment Station.

Worrall, J.G. 1980. The impact of environment on cambial growth, p. 127-142. *In* C. H. H. Little (ed.), *Control of Shoot Growth in Trees*. Proceedings of IUFRO Working Party on Xylem Physiology and Shoot Growth Physiology. Fredericton, New Brunswick, Canada.

Zahner, R. 1962. Terminal growth and wood formation by juvenile loblolly pine under two soil moisture regimes. *Forest Sci.* 8(4), 345-352.

SIMULATING FOREST ECOSYSTEM RESPONSES TO EXPECTED CLIMATE CHANGE IN EASTERN NORTH AMERICA: APPLICATIONS TO DECISION MAKING IN THE FOREST INDUSTRY

by Allen M. Solomon and Darrell C. West

ABSTRACT

The effects of rising levels of atmospheric carbon dioxide on climate are debatable. However, even if the climate change projections from general circulation models of the atmosphere prove correct, responses by forests on specific small tracts of land are not intuitively obvious. Standard regression techniques may be misleading or invalid for predictions of future structural or compositional responses; the ecological systems are simply too complex, and combinations of future climatic variables are too different from those of today. Instead, the problem may be soluble with a modeling approach based on documented differences in response by individual tree species to separate and combined environmental variables. One such approach, under development for several years at Oak Ridge National Laboratory and elsewhere, is used to show how the individual natural histories of tree species (warmth and moisture requirements, shade and root-competition tolerance, regeneration and mortality characteristics) can be interfaced with intra-annual variations and shifting long-term means in climate variables. The resulting model can project long-term forest ecosystem behavior under expected future conditions that may be quite different from those found on the modern landscape. The model's use is illustrated by applying it to land transaction decisions in which long-term considerations are required.

The increasing buildup of carbon dioxide (CO_2) in the atmosphere has been measured since 1958 (Keeling et al., 1982). Its preindustrial value of about 270 to 280 parts per million (ppm) (Wigley, 1983) has increased to over 345 ppm today (Gammon et al., 1984). The effects of rising CO_2 on climate are not clear, but general circulation models (GCMs) of the atmosphere indicate global warming should result, with warming greatest at the poles and least at the equator (Manabe and Wetherald, 1980). The magnitude of warming is expected to be greater in the winter than in the summer (Manabe and Stouffer, 1980), although winter warming may begin more slowly because it is controlled by slowly changing ocean surface temperatures (Schneider and Thompson, 1981). Interannual temperature variation in North America could decrease in most seasons if differences between present and future climates are similar to differences between coldest and hottest years in instrumental weather records (Palutikof et al., 1984). North American precipitation changes suggested by GCMs and by hot-year/-cold-year differences are most ambiguous. The changes may include less rain in autumn, more in winter (Palutikof et al., 1984), and greater annual rainfall. The result could be to increase evaporation, resulting in more summer drought in continental areas (Manabe et al., 1981; Wigley et al., 1980) and perhaps increased soil moisture in certain coastal areas (Wigley et al., 1980; Mitchell, 1983).

Even if the climate change projects are assumed to be valid, predicting responses of forest communities to the changes is not easy. Although trends across large regions can be predicted (for example, northward migration of forest communities and invasion of tundra by boreal conifers with warming), industry decision makers are not responsible for managing forests that constitute a larger region. Instead, because of the large number of small forest product industries, managers need to know what will occur on the relatively few commercial forest tracts for which they are now responsible or which they may acquire in the future. At the tract level, ecological systems are simply too complex to predict future structural or compositional variables, particularly among unmanaged forests. Unmanaged stands usually contain multiple species, each adapted differently to surviving environmental extremes and to competing for the limited sunlight, nutrients, and water required for successful growth and reproduction. Each developmental stage of tree growth (seedling, sapling, canopy tree) of each species also can respond differently to stresses of competition and of environmental limits.

Perhaps the most valid approach to resolving this temporal complexity is to construct and apply a computer model to simulate the interactions among species that follow from their individual silvical characteristics and are regulated by external environmental stresses. This approach has recently been applied to estimating effects on forests of future CO_2-induced climate changes (Solomon et al., 1984). This paper describes the model concept and its potential for using expected CO_2-induced climate changes to predict site-specific forest responses of direct significance to the forest industry. Specifically, the model assumes that industry decisions of relevance will involve specific tracts of commercial forest, with predictive information required on the ecological advisability of future retention, divestiture, or acquisition of those tracts.

MODEL CONCEPT AND STRUCTURE

The model FORENA (Forests of Eastern North America), under development at Oak Ridge National Laboratory for the past 10 years, will continue to undergo revision for some time. The current version is derived from FOREST (Forests of East Tennessee) (Shugart and West, 1977; Shugart, 1984), which in turn was modified from JABOWA (Janak Botkin Wallis) (Botkin et al., 1972). The model consists of constants (species parameters) that characterize silvical features of each species (Table 1) and variable external forcing factors that act on the species parameters. The spatial unit of the model is a 1/12-hectare (1/5-acre) area in an otherwise closed forest canopy. Plot size (Shugart and West, 1979) and simulated ecological features were chosen to mimic the dynamics of gap-phase regeneration in forests (Watt, 1947; Bray, 1956; Runkle, 1981; Shugart, 1984).

Tree dynamics are simulated with annual time steps. Simulated tree dynamics (intrinsic processes) involve effects of shading and crowding on growth (annual diameter and height increment of each tree on the plot), birth (additional of new seedlings to the plot), and death (loss of trees from the plot), all of which differ among species. Smaller time units (months, days) are used in calculating the external forcing factors that affect annual growth. External forcing factors (extrinsic variables) that regulate interactions among growth, birth, and death include cumulative summer warmth, extreme winter cold, and growing-season soil moisture, with differential effects on each species.

Intrinsic Processes

The intrinsic processes of forests were simulated as follows.

Growth

The basic concept implemented in the growth subroutine is that nothing can increase the defined optimum growth of a species, but various limited factors can each reduce annual growth from that optimum. Under the best growth conditions (freedom from crowding, shading, and temperature and water stress), the cumulative change in annual growth of diameter of many tree species forms

Table 1. Species parameters used in current version of forest-stand simulation model. Genus and species are the scientific names. For each species, Max age is the greatest age it can reach, Max diam is its maximum diameter in meters, Max ht is its maximum height in meters, Min GDD is its minimum growing degree-day tolerance, Max GDD is its maximum growing degree-day tolerance, Min Jan. is its minimum January temperature tolerance, % dry days is its maximum tolerance to drought in proportion of growing season days with soil moisture below the wilting point, Shade tol indicates its light-response equation (1 = shade tolerant; 2 = shade intolerant), Sprouting no. is the number of root sprouts added to the plot upon death of the tree, Sprouting Min is the minimum age required to sprout, Sprouting Max is the maximum age after which the tree cannot sprout, Birth switches are a series of seedling requirements that determine eligibility to grow each year (true [T] or false [F]) in sequence: requires mineral soil, requires leaf litter, tolerates no summer drought, is preferred wildlife food, requires full sunlight on the plot. The first two switches were not used in the simulation experiment described below. Max diam increm is the maximum annual growth possible, derived from maximum age and diameter, and used in the age-independent mortality routine.

Genus	Species	Common name	Max age (years)	Max diam (m)	Max ht (m)	Min GDD (°C)	Max GDD (°C)	Min Jan. (°C)	% Dry days	Shade tol	Sprouting No.	Sprouting Min	Sprouting Max	Birth switches	Max diam increm (cm)
Abies	*balsamea*	Balsam fir	200	0.5	15	560	2386	-25	0.165	1	0	0	0	FFTFF	0.41
Abies	*fraseri*	Fraser fir	200	1	35	2663	2763	-7	0.025	2	0	0	0	FFTFF	1.58
Acer	*rubrum*	Red maple	150	1	30	1260	6600	-18	0.230	1	3	12	150	FTTTF	1.77
Acer	*saccharinum*	Silver maple	125	1.2	30	1600	4700	-12	0.268	2	1	6	50	FTTFT	3.02
Acer	*saccharum*	Sugar maple	300	1.5	30	1222	3100	-18	0.193	1	3	12	80	FFTTF	1.26
Aesculus	*octandra*	Yellow buckeye	100	1	30	2660	3571	-1	0.175	1	1	12	100	TFTTF	2.71
Betula	*alleghaniensis*	Yellow birch	300	0.5	25	1100	2700	-18	0.343	1	3	12	100	FFTTF	0.80
Betula	*lenta*	Sweet birch	265	0.75	21	1402	3066	-2	0.177	1	3	12	100	FTTTF	0.95
Betula	*papyrifera*	Paper birch	140	1	25	484	2036	-28	0.347	2	1	10	30	FTTTF	2.30
Betula	*populifolia*	Gray birch	250	0.25	10	1007	2880	-11	0.292	2	1	10	30	FTFTF	0.44
Carpinus	*caroliniana*	American hornbeam	150	0.25	10	1344	6011	-12	0.382	1	2	6	70	FFFFF	1.12
Carya	*cordiformis*	Bitternut hickory	300	1	30	1910	5076	-12	0.320	1	1	12	200	TFTFF	0.70
Carya	*glabra*	Pignut hickory	300	1	30	1910	6960	-7	0.294	1	2	12	200	TFTFF	0.70
Carya	*laciniosa*	Shellbark hickory	300	1	30	2493	4615	-4	0.254	1	2	12	90	TFTFF	0.71
Carya	*ovata*	Shagbark hickory	275	1	30	1670	5500	-7	0.389	1	1	12	200	TFTFF	0.77
Carya	*texana*	Black hickory	300	1	30	2660	5076	-1	0.478	1	2	12	110	TFTFF	0.71
Carya	*tomentosa*	Mockernut hickory	300	1	28	1910	5993	-4	0.385	1	1	12	200	TFTFF	0.53
Castanea	*dentata*	American chestnut	300	1.5	35	1910	4571	-2	0.300	1	3	12	200	TFFFF	1.82
Celtis	*laevigata*	Sugar hackberry	200	0.75	30	2660	6960	-1	0.300	1	2	6	115	TFFFF	0.80
Cornus	*florida*	Flowering dogwood	100	0.25	10	1910	5993	-4	0.387	1	3	12	200	FFTFF	0.86
Fagus	*grandifolia*	American beech	366	1	30	1326	5537	-12	0.331	1	2	6	300	FFTFF	0.77
Fraxinus	*americana*	White ash	300	1	30	1398	5993	-12	0.414	1	2	6	20	FTTTF	1.25
Fraxinus	*nigra*	Black ash	300	1	25	1000	2261	-18	0.022	2	2	6	20	FFFTT	0.89
Fraxinus	*pennsylvanica*	Green ash	150	1	30	1050	5482	-23	0.114	2	1	6	50	FTFFF	1.43
Fraxinus	*quadrangulata*	Blue ash	300	1	30	2246	3733	-4	0.241	2	2	6	20	FTFTF	0.71
Ilex	*opaca*	American holly	200	0.75	20	2660	5993	2	0.360	1	1	6	40	TFTFF	1.08
Juglans	*cinerea*	Butternut	100	1	30	1870	3267	-12	0.333	1	1	6	12	FTTFF	3.26
Juglans	*nigra*	Black walnut	250	1.5	35	1910	4571	-8	0.429	2	1	5	20	TFTFF	1.71
Juniperus	*virginiana*	Eastern redcedar	300	0.75	20	1721	5537	-10	0.397	2	0	0	0	FTFFT	0.71
Larix	*laricina*	Tamarack	335	0.75	20	280	2660	-29	0.267	2	0	0	0	FTTTT	0.64
Liquidambar	*styraciflua*	Sweetgum	250	1.25	35	2660	5993	-1	0.423	2	2	12	80	FFFTF	1.05
Liriodendron	*tulipifera*	Yellow-poplar	300	1.5	35	1910	5993	-2	0.327	2	2	12	200	FTFTF	2.11
Nyssa	*sylvatica*	Black gum	300	1	30	1910	6960	-2	0.301	3	1	80	200	FTFFF	0.88

Table 1. (continued).

Genus	Species	Common name	Max age (years)	Max diam (m)	Max ht (m)	Min GDD (°C)	Max GDD (°C)	Min Jan. (°C)	% Dry days	Shade tol	Sprouting No.	Sprouting Min	Sprouting Max	Birth switches	Max diam increm (cm)
Ostrya	*virginiana*	Eastern hophornbeam	100	0.5	15	1278	5556	-18	0.360	1	2	6	20	FTFFF	1.06
Picea	*glauca*	White spruce	200	0.5	25	280	1911	-30	0.309	1	0	0	0	FFFFF	0.53
Picea	*mariana*	Black spruce	250	0.4	20	247	1911	-30	0.170	1	1	10	20	FFTFF	0.34
Picea	*rubens*	Red spruce	400	1	30	1247	2462	-12	0.237	2	0	0	0	FTTFF	0.60
Pinus	*banksiana*	Jack pine	150	0.5	25	830	2216	-30	0.511	2	0	0	0	FFTTF	0.60
Pinus	*echinata*	Shortleaf pine	300	1	30	2660	5076	-1	0.423	2	2	6	20	FTFFF	0.70
Pinus	*resinosa*	Red pine	310	0.75	25	1100	2035	-20	0.385	2	0	0	0	FTTFF	0.56
Pinus	*rigida*	Pitch pine	200	0.75	20	1940	3100	-7	0.307	2	2	6	20	FTFFF	1.06
Pinus	*strobus*	White pine	450	1.5	35	1100	3165	-20	0.267	2	0	0	0	FTTFF	0.69
Pinus	*taeda*	Loblolly pine	350	1	35	3165	5993	4	0.360	2	0	0	0	FTFFF	0.75
Pinus	*virginiana*	Virginia pine	250	0.5	15	2660	3671	-3	0.226	2	0	0	0	FTFFT	0.63
Platanus	*occidentalis*	Sycamore	500	1.75	35	1926	5482	-7	0.120	2	2	6	50	FTTFT	1.28
Populus	*balsamifera*	Balsam poplar	200	0.75	25	555	2491	-30	0.267	2	3	20	200	FTTTF	1.33
Populus	*grandidentata*	Bigtooth aspen	70	0.75	25	1100	3169	-18	0.267	2	2	2	70	FTTTT	3.89
Populus	*tremuloides*	Trembling aspen	200	0.75	22	743	2461	-30	0.267	2	3	20	150	TTTTT	1.20
Prunus	*serotina*	Black cherry	200	1	30	2132	5993	-10	0.387	2	4	12	200	FFFTF	1.07
Quercus	*alba*	White oak	400	1	35	1721	5537	-12	0.406	1	2	12	40	TFFFF	1.07
Quercus	*borealis*	Gray oak	250	0.5	25	1100	2200	-17	0.258	1	2	12	40	TFFFF	0.54
Quercus	*coccinea*	Scarlet oak	400	0.75	25	2038	4571	-7	0.286	1	2	12	80	TFFFF	0.68
Quercus	*ellipsoidalis*	Northern pin-oak	200	0.75	25	1279	2234	-15	0.318	2	3	12	200	FTFTF	0.95
Quercus	*falcata*	Southern red-oak	400	1	35	2660	5993	2	0.423	1	2	12	30	TFFFF	0.94
Quercus	*lyrata*	Overcup oak	250	0.8	25	2926	5315	4	0.031	2	1	6	12	TTTTF	1.55
Quercus	*macrocarpa*	Bur oak	300	0.8	25	1157	5153	-20	0.487	1	2	10	30	TTFFF	1.23
Quercus	*marilandica*	Blackjack oak	400	0.5	15	2493	5537	-1	.422	2	2	12	40	FFFFF	0.43
Quercus	*muehlenbergii*	Chinkapin oak	300	1	30	1958	4849	-7	.383	2	2	10	30	TTFFF	1.05
Quercus	*nuttallii*	Nuttall oak	250	0.75	25	3371	5260	4	.030	2	1	6	12	TFTTT	1.21
Quercus	*palustris*	Pin oak	200	0.75	25	2217	5153	-6	.013	2	3	10	100	TTTFT	1.87
Quercus	*prinus*	Chestnut oak	267	1	30	1910	4110	-7	.285	1	2	12	40	TFFFF	1.43
Quercus	*rubra*	Northern red-oak	400	1	30	2000	4571	-17	.225	1	2	12	40	FTFFF	1.06
Quercus	*shumardii*	Shumard oak	300	1	35	2393	5993	-3	.484	2	2	12	160	FTFFF	1.40
Quercus	*stellata*	Post oak	400	0.75	25	2660	5993	-4	.555	2	2	12	30	TFFFF	0.54
Quercus	*velutina*	Black oak	300	1	30	1810	5076	-10	.360	1	2	12	40	FTFFF	1.24
Quercus	*virginiana*	Live oak	300	1.5	20	4849	6674	7	.512	1	3	10	150	TTTFF	2.08
Thuja	*occidentalis*	Northern whitecedar	400	1	35	1000	2188	-20	.350	1	1	5	400	TTTTF	0.80
Tilia	*americana*	American basswood	140	1	30	1400	3137	-17	.270	1	3	12	80	FFFTF	2.73
Tilia	*heterophylla*	White basswood	150	1	30	2660	4571	-1	.211	1	3	12	80	FTTFF	1.78
Tsuga	*canadensis*	Eastern hemlock	650	1.5	35	1324	3800	-12	.288	1	0	0	0	TFTTF	0.57
Ulmus	*alata*	Winged elm	125	0.75	20	2660	5993	2	.345	2	1	6	110	FFFFF	1.54
Ulmus	*americana*	American elm	300	0.8	25	1204	6960	-20	.457	2	2	6	240	FFTFF	1.95

an S-shape, being small at first, reaching a maximum before middle age, and decreasing thereafter (Figure 1a). The model utilizes a similar growth curve (Figure 2a), defined separately for each species from its maximum diameter in closed forest stands and its maximum known age (*see*, for example, Fowells, 1965; Harlow et al., 1979). The "optimum" growth curve is arbitrarily defined so that a tree reaches two-thirds of its maximum diameter at one-half of its maximum age (Botkin et al., 1972). The differing maximum ages and diameters of each species generate species-specific sizes at any given age (Figure 2a). Optimum growth may actually be achieved by trees growing in plantations designed to minimize effects of competition for sunlight and nutrients if weather during growth is also optimal, but it is rarely, if ever, accomplished by trees in unmanaged forest stands.

Optimum growth is reduced by shading. Among other factors, photosynthesis determines annual growth (Kramer and Kozlowski, 1979). Photosynthesis, in turn, depends on the interaction between irradiance and shade tolerance (*see*, for example, Kramer and Decker, 1944; Figure 1b). Irradiance decreases from the canopy top to the forest floor. The model tracks the height of the leaf area of each simulated tree (a function of diameter and empirical constants from Sollins et al., 1973), averaging the cumulative leaf area across the plot at vertical intervals of 0.1 m (0.33 feet). The relationship between irradiance and growth is modeled by using field-measured light-extinction equations (Monsi et al., 1973). The calculated leaf areas above the top of each tree on the plot are used to estimate the proportion of full sunlight on each tree. Suboptimal light proportionately reduces the optimum growth of all trees on the plot by height, with separate response equations for shade-tolerant and shade-intolerant species (Figure 2c). Shade tolerance of each species is taken from the data summarized by Fowells (1965).

A second intrinsic limit to growth is nutrient availability, itself a complex and dynamic series of processes. The model simplifies the nutrient limits with an indirectly related variable, tree density. Growth of all trees on the plot decreases linearly from their optima as biomass on the stand increases (Figure 2d). The linear form of the stand's density response is, at best, a first approximation that appears to work in many situations but is currently being replaced with nutrient availability and response routines by individual species (Weinstein et al., 1982; Pastor and Post, 1984).

Figure 1

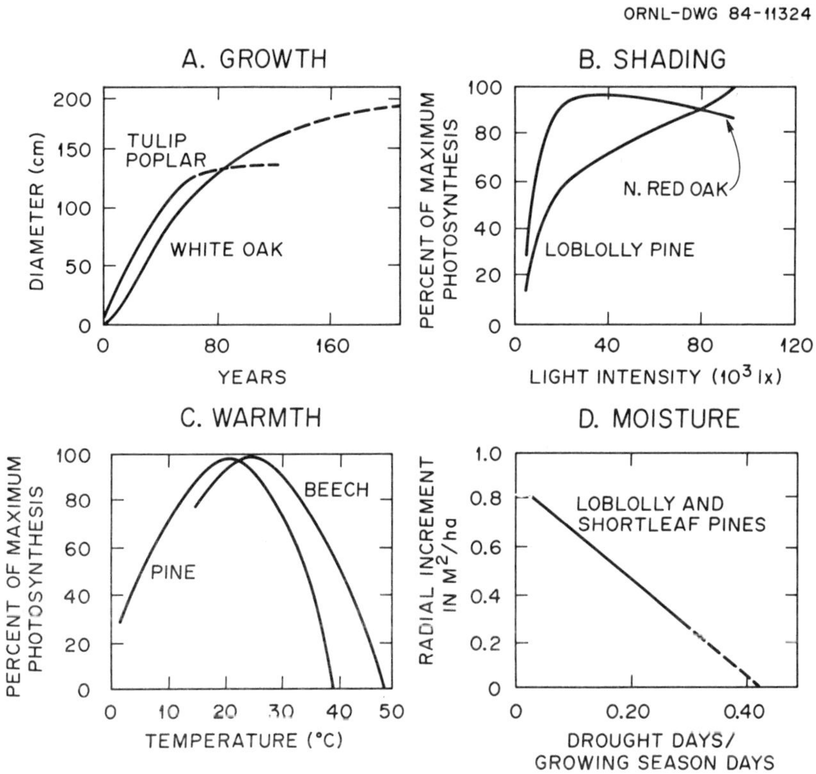

Tree growth responses to external variables defined experimentally and applied to the simulation mode (Figure 2). (A) Diameter growth of white oak and tulip (yellow) poplar as a function of age, with minimal competition for light, water, and nutrients near the optimum portion of their geographical ranges (from data of Doyle, 1983). (B) Growth response to shading in red oak and loblolly pine (redrawn from Kramer and Decker, 1944). (C) Growth response to temperature in European species of beech and pine (redrawn from Larcher, 1969). (D) Growth response to days in which soil moisture is below the wilting point as a proportion of annual number of growing season (GS) days in (combined) loblolly and shortleaf pines (redrawn from Bassett, 1964).

Figure 2

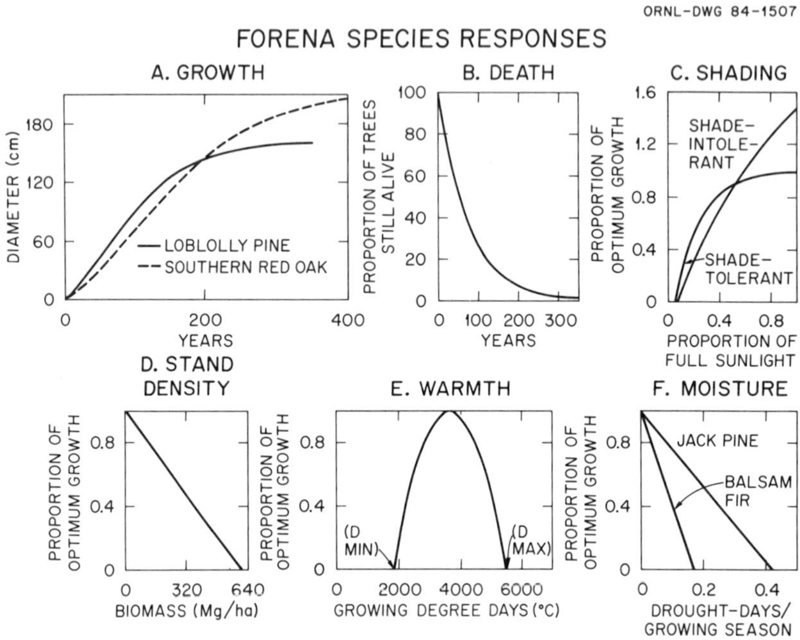

Tree growth responses to external and biological variables as made formal in the simulation model FORENA, from Solomon and Tharp (1984). (A) Diameter growth of two tree species as a function of age in the absence of competition for light, water, and nutrients at the optimum portion of their geographical ranges. (B) Mortality probabilities of trees as a function of tree age and the maximum age (AGEMAX) the tree species can achieve; mortality of loblolly pine is illustrated. (C) Growth response to shading in shade-tolerant and shade-intolerant species (see Table 1 for tolerance classification of species; see Shugart and West, 1977, for the separate growth equations that generate the response curves). (D) Growth response to stand density by all species. (E) Growth response to annual growing degree days. "D MIN" and "D MAX" are the lowest and highest degree-day values, respectively, within which the species can grow. Degree-day requirements of sycamore are exemplified. (F) Growth response to days in which soil moisture is below the wilting point as a proportion of annual number of growing season (GS) days. "Drought-day" requirements of jack pine and balsam fir are exemplified.

Death

There are two independent death processes in the model. First is death resulting from senescence. Several effects of aging--for example, decreasing radial growth·capacity (Bormann, 1965), decreasing crown-to-stem ratios (Jacobs, 1955), and increasing root mortality (Copeland, 1952), coupled with decreasing root-regeneration capacity (Redmond, 1959) and decreasing resistance to pathogen (Kozlowski, 1969)--increase the probability of death with age, although empirical data that define these probabilities has not been found. Instead, the probability of death is assumed to increase logarithmically with age (Figure 2b), so that only 1 of every 100 trees reaches the maximum age assigned to the species.

The second source of mortality depends less on tree age. Many factors can slow or stop the annual growth of a vigorous tree, such as late-spring frosts (Tyrone and True, 1968), too much or too little soil water (White, 1969; Hicks, 1980), fire or lightning damage (Schowalter et al., 1981), nutrient deficiencies (White, 1974; Mattson, 1980), and defoliation by insects (Nichols, 1968; Mattson and Addy, 1975). Slow growth at any age increases the probability of tree death by reducing the photosynthesis of food for metabolism of living cells (Kramer and Kozlowski, 1979) and by decreasing resistance to pathogen attacks (Hepting, 1963; Schmidt, 1978). Nichols (1968) found that at least two or three consecutive years of slow growth are required to kill mature trees.

From scattered data on maximum and minimum annual diameter increments among forest trees (Fowells, 1965; Harlow et al., 1979), slow growth is defined in the model as 1/10 of the maximum possible annual growth each species can achieve. When a simulated tree undergoes such slow growth in two consecutive years, the modeled probability of death increases by 0.308. This value is similar to the one determined by Nichols (1968) for a comparable mortality variable after three consecutive years of slow growth. The result of this routine is that shading, crowding, and other modeled stresses can cause tree mortality, particularly in old and very young trees and in trees that grow rapidly (that is, low maximum age and large maximum diameter) and, therefore, possess large minimum increments of annual diameter (Table 1). Conversely, stress that reduces growth favors young, mature trees and trees that can grow slowly (that is, high maximum age and moderate maximum diameter).

Birth

When a tree dies in a closed-canopy forest, a canopy gap is created in which sunlight allows rapid growth of suppressed seedlings and saplings (*see*, for example, Shugart, 1984). The model mimics this process by annually "planting" a random number of saplings, each 2.5 cm in diameter or about 1 inch. The species of trees planted are randomly chosen after the requirements for fresh mineral soil, leaf litter, and sunlight (Table 1) of each seedling species are compared with those conditions on the plot. In addition, a random number of saplings of species known to sprout from rootstocks are added to the plot if parent trees die at sizes with specific ranges of diameter (Table 1). If sunlight is inadequate for vigorous growth, the slow-growth mortality routine eventually kills the new saplings. However, if a tree's death creates a canopy gap, the saplings present begin to grow rapidly. The tree or trees that eventually reach the canopy do so because, among the random selection of saplings present, they are of the proper age and species to grow the most rapidly under the shading conditions created by the remaining trees.

When the model is run, the foregoing routines reproduce a temporal sequence of differing tree species, stem sizes, and age classes that might be observed in an ideal environment of optimum constant temperature and precipitation. In reality, environmental variations act directly on tree composition and numbers through infrequent extremes--for example, catastrophic fires, floods, windstorms, frosts, and droughts (not simulated in this version of the model). Environmental variations also indirectly regulate the tree composition and numbers on such plots by frequently shifting competitive advantages among the species. To mimic these processes, the model simulates effects of temperature and precipitation variations as further reductions in optimum annual diameter increments.

Extrinsic Variables

Each climatic effect routing contains two components: climatic force and tree responses. Climate forcing is in terms of variations (that is, random values drawn from normal frequency distributions defined by means and standard deviations). The means define the center of the range of values within which the forcing variable will be found, but random chance determines what the value of the variable will be. The normal distribution may be a weak substitute for certain climatic variance functions (for example, skewed distributions of desert precipitation or tundra temperatures), but it is an excellent approximation compared with the assumption that climatic variance does not affect tree growth (that is, JABOWA, in Botkin et al., 1972; FORTNITE, in Aber and Melillo, 1982).

Temperature

Warmth controls modeled tree growth in two ways. First, climatic forcing is provided by the annual accumulation of heat units (growing degree days, or GDD) above a temperature threshold. GDD values are calculated from random average monthly temperatures above 5.5°C (42°F) (*see*, for example, Lindsey and Newman, 1956). The parabolic form of the temperature-photosynthesis response curve (Larcher, 1969; Figure 1c) provides the GDD force and tree response function for the model (Figure 2e), although no direct evidence exists that the GDD-photosynthesis relationship is interchangeable with the temperature-photosynthesis relationship. Minimum and maximum GDD tolerances are determined from measured GDD values at the northern and southern edges, respectively, of each species' geographical range. Loss of potential growth is zero when the annual random GDD value matches that at the center of a species' range and is up to 100% growth loss when the GDD value falls outside those in the species' range. Effects of irregular, extreme winter temperatures are calculated separately. The random value for January temperature, generated as one of the 12 values needed to calculate GDDs, is compared with the lowest average January temperature recorded (Cohen, 1973; Hare and Thomas, 1979) within the geographical range of each species (Little, 1971 and 1977). If the random January forcing temperature is the lower of the two values, tree growth is zero for the year to mimic loss of leaf buds (*see*, for example, Tyrone and True, 1968). Also, seedlings are not eligible to be planted that year to mimic loss of flower buds and, thus, seeds (Sakai and Weiser, 1973). Finally, two or more such "no growth" responses in consecutive years invoke the age-independent mortality routine, increasing the probability of mortality.

Moisture

Precipitation affects annual growth in diameter through soil moisture availability (Bassett, 1964; Zahner and Donnelly, 1967. The model uses Thornthwaite equations (Thornthwaite and Mather, 1957) to determine the monthly water content of a soil of specified depth (rooting zone, 50 to 100 cm, or 20 to 40 inches), field capacity (-0.1 bar, soil column full), and wilting point (-15 bars, soil column empty). The model adds the random value for monthly precipitation and subtracts the monthly water loss from evapotranspiration (from random monthly temperature and latitude) and runoff (excess over field capacity). Although the model tracks water content of the soil during all months, it forces tree growth with the annual drought days (DDs), which include only growing-season (GS) days during which soil moisture is below the wilting point. The linear species response to annual DD values determined by Bassett (1964) (Figure 1d) is the basis for modeled species responses to DDs (Figure 2f). Annual tree growth decreases linearly from optimum, when there are no DDs, to the maximum DD value. This is determined for each species from western-range limits (Little, 1971 and 1977) superimposed on county-level maps of DDs on north-facing clay-loam slopes, clay-soil floodplains flooded for three-fourths of the GS, or sandy upland soils, whichever is appropriate for the species at the western edge of its range (Fowells, 1965; Harlow et al., 1979). The maps were calculated from data based on county-level monthly precipitation, temperature, and GS (Olson et al., 1980). Because the length of the GS decreases from south to north within a species' geographical range, the maximum DD constant for each species is the proportion DD/GS (Table 1). This value is multiplied by the specific GS values at the modeled site before each simulation.

The modeled effects of these extrinsic forces are chronic rather than catastrophic (that is, they do not kill trees directly). Rather, they shift competitive advantages among the species by changing the species-specific effects of intrinsic birth, growth, and mortality characteristics, reducing growth of some species while enhancing that of others, and thereby gradually changing stand structure and species composition. The way in which stand structure (diameter size classes) and composition will change with specified climate forcing is not intuitively obvious in the absence of such a model because the stand characteristics that will be most important will vary from place to place, depending on the site's characteristics, the stand's age, and the mix of species, their silvical features, and their ages. Yet, it is precisely this kind of information on commercial forests that is available to decision makers in the forest industry and it is chronic rather than catastrophic effects that commercial forests are likely to suffer due to CO_2-induced climate change.

There are several less-complicated interactive versions of the model. The operator can designate initial environmental conditions, presence of bare soil or of a stand with specific composition and stem diameters, and then view the model output at intervals while changing the conditions (that is, selectively logging, clearcutting, burning the plots). These versions might be particularly valuable in following the short-term implications of differing management strategies.

MODEL VALIDITY

The validity of the approach and of the specific model versions is difficult to demonstrate comprehensively. There is a continuum of both extrinsic forcing-value combinations (for example, temperature and precipitation) and of intrinsic measurements (stand structure, composition, and density) of forest responses. Earlier versions of the model were validated with available measurements of forest stands at the specific sites for which the model was tailored (*see*, for example, Botkin et al., 1972; Shugart and West, 1977; Aber et al., 1978; Tharp, 1978; Phipps, 1979).

Later research was aimed at using a single model format to replicate 10,000- to 20,000-year forest histories containing multiple types of forest ecosystems. The internal consistency of the model in different forest ecosystems was tested by simulating the vegetation indicated in fossil pollen records, based on climate variables extracted from the pollen records (Solomon et al., 1980 and 1981; Solomon and Tharp, 1984). The validity of the simulations was independently measured when the model reproduced the specific forest responses seen in fossil pollen records after being constrained with climate variables extracted from independent pollen or nonpollen paleoenvironmental indicators (Solomon, 1982; Solomon and Shugart, 1984).

After model modifications indicated by these latter tests had been implemented, the simulation of the spatial variety of today's upland forest ecosystems in eastern North America (Solomon et al., 1984) was attempted. The first 400 years of the 1000-year simulations were constrained by the climatic means and variances of the 1931 to 1960 "normal" period, in part to validate the model. Considering that the simulations included 72 tree species at 21 sites, there were over 1500 opportunities for model responses to be "wrong" (species growing where or when they should not; not growing where or when they should; growing in too great or too little abundance). Fewer than 150 such wrong responses were recorded when simulated trees first were classed as dominant, subdominant, infrequent, or absent at each site, and then were compared with species in the same classes (Kuchler, 1964; Rowe, 1972) expected in the actual forests at each simulation site.

Many of these disparities may be reduced by the use of improved data to more carefully draw the parameters of the species in the model (for example, northern white cedar, too plentiful at four sites because of incorrect diameter and height parameters, thereby making associated species too sparse at the same sites; Solomon et al., 1984). Other inconsistencies appear to be related to the single-species characterization of multiple ecotypes (for example, beech, as described by Camp, 1951; and hemlock, as described by Kessell, 1979; *see also* Solomon and Tharp, 1984; Solomon et al., 1984). An unknown portion of the remaining wrong responses probably result from environmental forces (some of which are available as forcing variables in other versions of the model) not simulated, such as realistic frequencies of catastrophic disturbances, including fire, windstorm, disease, and insect epidemics; nutrient cycling and local soil variations; air pollutant stresses and gradual warming of the past 100 years.

Empirical and statistical approaches (*see*, for example, U.S. Department of Agriculture [USDA], 1982) are probably more effective than the foregoing model for projecting short-term (±30 years) change in large regions, changes similar to those recorded in recent history. However, with the notable exception of one phenomenon that neither method can assess (see below), the simulation model seems more applicable than statistical estimates for projecting long-term, site-specific responses of commercial forest tracts to environmental variations attributable to increasing atmospheric CO_2. Monthly and seasonal temperature and precipitation, and their variances, as well as growing season length and seasonal temperature extremes, are forcing variables in the model that are expected to change in the future. Annual responses to many of these forces are realistically simulated, while certain unmodeled seasonal responses (growth of tree diameter as a function, primarily, of spring moisture or of winter temperatures; *see*, for example, Fritts, 1976) can easily be incorporated in a manner similar to the treatment of winter low temperature extremes, already in the model.

The exception to expectations of reliable model application involves the potential effects of direct CO_2 effects, as outlined by Leverenz et al. (this volume). In another paper, Kramer and Sionit (this volume) discuss the possible characteristics of CO_2 fertilization and describe the meager knowledge base supporting the current estimates of future effects. Model routines can be designed within the current framework for treating overall CO_2 enhancement of photosynthesis (and of nutrient requirements from available pools), increases of photosynthesis in shade (and of shading from more and thicker leaves), greater water-use efficiency (and water use caused by increased evapotranspiration), decreased photorespiration (and increased dark respiration from warming), and related phenomena. However, the routines would, first, require response functions for whole trees (as opposed to single leaves or branches) to changes in photosynthesis, water use, and photosynthetic efficiency in shade, and second, at least a notion of the systematic differences in response among tree species, such as those indicated by the geography of species range. Although a marginal amount of the information for the first requirement above is available to treat seedlings, none has yet been generated for mature trees, and the data needed to develop the second requirement do not appear to exist. Therefore, for the present, the current model structure is considered to be the most reliable, scientifically justifiable vehicle for projecting effects of gradually increasing atmospheric CO_2, with the caveat that new scientific information on direct CO_2 effects probably will both decrease the unknowns and enhance the validity of this model approach.

MODEL APPROACH APPLICATIONS

How might the simulation model approach be integrated into forest industry decision-making processes? Several kinds of problems of various appropriateness suggest themselves. For example, an interactive version of the model might be used to project the short-term results of different management options, applied to specific forested sites, under projected CO_2-induced climate changes. Below, the model is used to provide information to be combined with economic data in decisions on forestland transactions. The three situations that follow are simplified, but they exemplify the potential use, as well as the strengths and weaknesses, of the approach. The hypothetical question is whether to sell or retain current holdings or to acquire additional tracts of forestland in each of three regions.

The CO_2 question is irrelevant to the short-term considerations involving, for example, recent forest productivity, age structure of stands, timber accessibility, or current product demand, because resulting climate changes will be barely (if at all) measurable in the next 25 or 30 years (Wigley et al., 1980), let alone a significant factor in forest growth. The CO_2 issue only becomes relevant in long-term planning, in concert with projections of changing tax structure, land values, and so on.

Model simulations of the potential growth of the 72 species in Table 1 for 1000 years at 21 sites were attempted (Solomon et al.,

1984). Among them, sites in central Ontario, the western upper peninsula of Michigan, and middle-east Tennessee were chosen to answer the hypothetical question posed above for the present exercise.

Simulation Conditions

During the first 400 years of the simulations, current climate means and variances prevail to allow trees to seed into and grow on the plots (the model could have been constrained with diameters, heights, and ages of trees actually growing on specific plots). Then, CO_2 induces climate change, with the timing of change based on estimates by Perry (1982) and the amount of change based on experiments with a general circulation model of the atmosphere at doubled and quadrupled CO_2 (Mitchell, 1983; Michell and Lupton, 1984). In this scenario, CO_2 doubles between 400 and 500 simulated years, including warming (+5.0°C or 9.0°F winter, +2.5°C or +4.5°F summer in Ontario; +3.5°C or +6.5°F winter, +2.5°C or +4.5°F summer in Michigan; +2.0°C or +3.5°F winter, +3.0°C or 5.5°F summer in Tennessee), with little absolute change in seasonal precipitation (no change in Ontario and Michigan; -1.0 cm or -0.5 inch per month winter, +1.0 cm or +0.5 inch per month summer in Tennessee). Between 500 and 700 years, CO_2 doubles again (quadrupling today's concentration), inducing additional warming (+4.0°C or +7.0°F winter, +2.5°C or +4.5°F summer in Ontario and Michigan; +3.0°C or 5.5°F winter, +2.0°C or +3.5°F summer in Tennessee) with no precipitation change (Ontario) or else with slightly less precipitation (-3.0 cm or -1.2 inches per month winter, no change summer in Michigan; -5.0 cm or -2.0 inches per month winter, -1.0 cm or -0.5 inch per month summer in Tennessee). From 700 to 1000 years, climate mean values are stable at the quadrupled CO_2 level. Climate variance remains the same throughout the simulations.

The simulations were each reiterated 10 times and the results were averaged. The soil is a 100-cm-deep (40-inch-deep) silt-loam with a wilting point of 18.5% and field capacity of 35.5%. The soil moisture capacities could have been tailored to conditions on specified land tracts.

The climate scenario seems to be relatively realistic through a doubling of CO_2 (to year 500) and increasingly speculative thereafter, with no evidence that quadrupling will eventually occur or that, if so, climate change will cease at a CO_2 quadrupling. However, the above sequence was used because simulated behavior of forest stands growing under continuously changing climate was the examination's focus. The reader is cautioned to consider the post-year-500 simulations as particularly hypothetical.

Western Ontario Simulation

The most relevant features of the simulations include the temporal changes on an average hectare of biomass by species or species group (Figure 3a); of biomass by ecosystem type (Figure 3b), of stems by ecosystem type (Figure 3c), and of biomass of an average tree (Figure 3d; total biomass divided by total number of stems). The simulations begin with initial dominance by paper birch, white spruce, and black spruce, with small amounts of balsam fir and jack and red pine (Figure 3a), all classified as boreal conifers (Figure 3b). After 150 years, white spruce alternates dominance with paper birch in a dynamic equilibrium.

Initial warming (years 400 to 500) has no clear impact on total stand biomass, because paper birch rapidly fills the canopy gaps resulting from sharp declines in biomass (about 15 Mg or metric tons per hectare) and stems (about 1300 per hectare) of white spruce and black spruce. Between years 500 and 575, stand biomass drops, and stems increase as a general dieback occurs among the boreal species. Thereafter, until the simulation ends at 1000 years, a northern hardwood forest gradually develops, dominated primarily by sugar maple with steadily increasing amounts of white pine. The result of the boreal species' replacement by hardwoods is an increase in stand biomass with higher frequencies of the larger hardwood species, a decrease in stem density with greater shading, and a concomitant increase in average tree size.

If the land were used for pulp production, as are many tracts in the region, simulations indicate reason for some concern about supply or production changes in spruce during the first 75 years of simulated warming and good reason for considering divestiture thereafter. Hardwood production is another matter because a mixture of species (red, bur, and Hill's oaks; green ash; northern white cedar; balsam poplar; and trembling aspen) comes and goes for about 200 years (about years 625 to 825) until the sugar maple and white pine finally assert dominance in diameters large enough to justify commercial development of the stands.

Northwest Michigan Simulation

During the first 400 years at this site, a succession of boreal and northern hardwood species develops. The first 100 years (Figure 4a) are dominated by paper birch, balsam fir, jack pine, and bal-

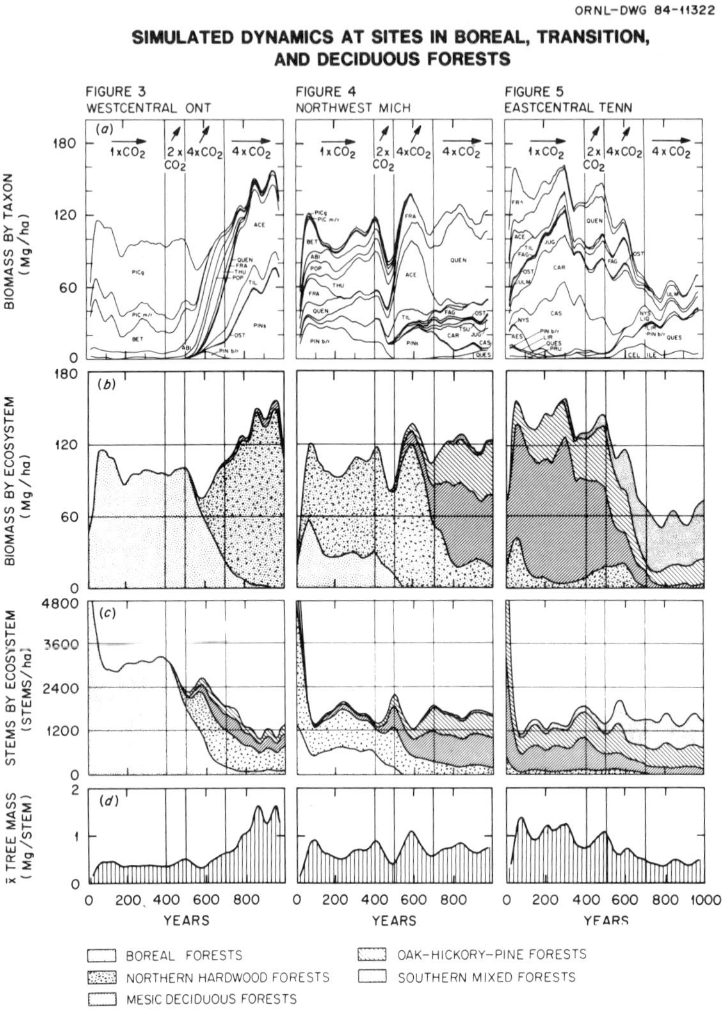

Figure 3

Simulated forest stand dynamics at a site in boreal forest for 1000 years; years 0 to 400 with stable climate of 1931 to 1960 (1X CO_2); years 400 to 500 with climate steadily changing to that induced by a CO_2 doubling at year 500 (2X CO_2); years 500 to 700 with climate steadily changing to that induced by a CO_2 quadrupling at year 700 (4X CO_2); and years 700 to 1000 with stable climate at the quadrupled CO_2 values. (a) Simulated biomass in megagrams or metric tons per hectare by taxon (taxon abbreviations listed in Figure 5 legend). (b) Simulated biomass in megagrams or metric tons per hectare by ecosystem type. (c) Simulated stem density (stems per hectare) by ecosystem type. (d) Simulated biomass by average (x) tree (megagrams or metric tons per stem) by ecosystem type.

Figure 4

Simulated forest stand dynamics at a site in mixed coniferous and deciduous forest, for 1000 years (see Figure 3 legend for climate change scenario). (a) Simulated biomass in megagrams or metric tons per hectare by taxon (taxon abbreviations listed in Figure 5 legend). (b) Simulated biomass in megagrams or metric tons per hectare by ecosystem type. (c) Simulated stem density (stems per hectare) by ecosystem type. (d) Simulated biomass of average (x) tree (megagrams or metric tons per stem) by ecosystem type.

Figure 5

Simulated forest stand dynamics at a site in mixed coniferous and deciduous forest, for 1000 years (see figure 3 legend for climate change scenario). (a) Simulated biomass in megagrams or metric tons per hectare by taxon. (b) Simulated biomass in megagrams or metric tons per hectare by ecosystem type. (c) Simulated stem density (stems per hectare) by ecosystem type. (d) Simulated biomass of average (x) tree (megagrams or metric tons per stem) by ecosystem type. Taxon abbreviations in Figures 3a, 4a, and 5a are:

ABI: Abies balsamea (balsam fir), A. fraseri (Fraser fir)
ACE: Acer saccharum (sugar maple), A. rubra (red maple), A. saccharinum (silver maple)
AES: Aesculus octandra (yellow buckeye)
BET: Betula lenta (sweet birch), B. papyrifera (paper birch), B. alleghaniensis (yellow hickory), B. populifolia (gray birch)
CAR: Carya cordiformis (bitternut hickory), C. tomentosa (mockernut hickory), C. glabra (pignut hickory), C. ovata (shagbark hickory), C. laciniosa (shellbark hickory), C. texana (black hickory)
CAS: Castanea dentata (American chestnut)
CEL: Celtis laevigata (sugarberry)
FAG: Fagus grandifolia (American beech)
FRA: Fraxinum pennsylvanica (green ash), F. americana (white ash), F. nigra (black ash), F. quadrangulata (blue ash)
JUG: Juglans cinerea (butternut), J. nigra (black walnut)
LIQ: Liquidambar styraciflua (sweet gum)
LIR: Liriodendron tulipifera (yellow poplar)
NYS: Nyssa sylvatica (blackgum)
OST: Ostrya virginiana (eastern hophornbeam), Carpinus caroliniana (American hornbeam)
PIC g: Picea glauca (white spruce)
PIC m/r: Picea mariana (black spruce), P. ruben (red spruce)
PIN b/r: Pinus banksiana (jack pine), P. resinosa (red pine), P. rigida (pitch pine), P. echinata (shortleaf pine), P. taeda (loblolly pine), P. virginiana (Virginia pine)
PIN s: Pinus strobus (white pine)
POP: Populus balsamifera (balsam poplar), P. grandidentata (bigleaf aspen), P. tremuloides (trembling aspen)
PRU: Prunus serotina (black cherry)
QUEN: (Northern) Quercus alba (white oak), Q. coccinea (scarlet oak), Q. prinus (chestnut oak), Q. rubra (northern red oak), Q. velutina (black oak), Q. macrocarpa (bur oak), Q. borealis (gray oak), Q. ellipsoidalis (northern pin oak)
QUES: (Southern) Quercus falcata (southern red oak), Q. lyrata (overcup oak), Q. marilandica (blackjack oak), Q. muehlenbergii (chinkapin oak), Q. nuttallii (Nuttall's oak), Q. palustris (pin oak), Q. shumardii (Shumard's oak), Q. stellata (post oak), Q. virginiana (live oak)
THU: Thuja occidentalis (northern white cedar), Juniperus virginiana (red cedar), Larix laricina (tamarack)
TIL: Tilia americana (American basswood), T. heterophylla (white basswood)
TSU: Tsuga canadensis (hemlock)
ULM: Ulmus americana (American elm), U. alata (winged elm)

sam poplar (all classified here as boreal species), and red ash, red pine, gray oak, and white pine (all classified as northern hardwoods). During the first 75 years, stem density declines from about 4200 per hectare to 1400 per hectare, primarily depicting the growth of individual trees and the resulting mortality from shading of northern hardwood seedlings (Figure 4b). From about 100 years to 400 years, sugar maple, white cedar, and white pine gradually assume greater importance at the expense of the birch, poplar, and ash. Notably, 100 to 300 stems per hectare of trees classified in mesic deciduous and oak-hickory-pine forests are present after about 150 years (Figure 4c), but they produce no significant biomass (that is, the saplings regularly die well before reaching maturity). The inverse relationship between stand biomass (Figure 4a and b) and stem number (Figure 4c) follows from the model's reproduction dynamic because loss of individual, large-biomass, mature trees generates many small-biomass seedlings.

An immediate dieback of most species dominates the initial 75 years of warming (years 400 to 475), including local extinction of boreal spruce and paper birch populations (Figure 4a and b). Decline of stand biomass is coupled with sharp increases in stem density and decreases in the average tree biomass. Inspection of the detailed simulation output indicates that the biomass and stem changes are due primarily to loss of saplings and older mature sawtimber, with greatly slowed growth among poletimber size classes. The quick response is not surprising, considering the ecotonal character of the communities. Many dominant tree species that grow at the southern edges of their respective geographical ranges die quickly, whereas those growing at their northern-range edges require considerable time to reach maturity.

The following 200 years (475 to 675) are characterized by the appearance, dominance, and decline of northern hardwood species (sugar maple, basswood, white pine, gray oak, and red and green ash; Figure 4a). A dynamic equilibrium appears to be reached by about year 800, with a forest dominated by oaks and hickories, with minor numbers of hemlock, beech, and ash.

The forecast for industry landholdings on such sites appears quite dim. Quick divestiture with the onset of warming would seem warranted, whether the holding industry is concerned with pulp or lumber. Early cessation of reproduction, large mortality losses in sawtimber, and reduced growth in poletimber would severely reduce commercial yields after the initial mortality was harvested. Replacement of conifer sawtimber size classes by sugar maple and white pine requires over a century (years 400 to 500) of warming, and the latter species decline quickly thereafter.

East-Central Tennessee

The forests that develop during the first 400 years are quite diverse, with 20 tree species contributing at least 1% of the stand's biomass (Solomon et al., 1984). Among them, American chestnut is the most important single species (averaging 25% of the biomass during the first 400 years; the decline in chestnut caused by blight was not simulated), accompanied by five hickory species (21%), four oak species (13%), and the white basswood (7%) and yellow buckeye (4%), which are characteristic of these mixed mesophytic hardwood forests (Braun, 1950).

In contrast to the two simulations previously described, no discernible effect on any of the simulated biotic variables (Figure 5a to d) emerges from the normal stand variability during initial climate change (years 400 to 500). This stability appears to result in part from the presence of many dominant species and in part from the plant geography of the many species simulated. For every individual that slows in growth, there is one of another species that begins to grow more rapidly. In addition, most of the species are growing more or less near the midpoint of their north-to-south ranges (simulated as optimum conditions). Thus, the initial warming produces little if any increase in frequency of annual temperature or precipitation values extreme enough to slow growth of most tree species.

The changes in climate following CO_2 doubling do cause reduced growth and local loss of all species except certain southern oaks (post oak, Shumard red oak, and southern red oak), American elm, loblolly pine, and sweetgum, resulting in a very species-depauperate forest. Although the stem density changes little during the last 500 simulated years, stand biomass and average tree size both decline to less than one-half of their values at doubled CO_2. Dynamic equilibrium is reached shortly after the climate stabilizes at year 700.

The simulations indicate that forest productivity responses to climate change, induced by at least a CO_2 doubling, may not need to be considered in decisions on land transactions at such sites. The primary changes in forest structure, density, and composition occur so late in the simulations that they are irrelevant to most land transaction decisions. Coincidentally, the climate scenario this far into the future is worth little in any predictive sense.

DISCUSSION

The three simulations described illustrate some of the generic strengths and weaknesses of the simulation approach when it is applied to decisions on landholdings that involve perceptible effects of CO_2-induced climate changes. On the positive side, the model generated certain information not otherwise currently available. For example, assuming that forest dynamics are controlled by the birth, growth, and death characteristics of the individual tree species available, the model documented the variably lagged responses of species composition, stand structure, and density by forest stands that may be expected from the hypothetical climate changes used. If valid, such information could provide the biotic component missing from any long-term economic analysis of industry holdings and choices.

Additionally, the simulations indicate how forest productivity would behave under gradually and continuously changing climate. The rate of seasonal temperature increase in the scenarios ranged from a minimum of 0.02°C or 0.035°F (summer in Ontario and winter in Tennessee for a CO_2 doubling) to a maximum of 0.05°C or 0.09°F (winter in Ontario for a CO_2 doubling) each year. None of the CO_2-doubling scenarios contained precipitation changes greater than 0.01 cm (0.005 inch) per month (summer and winter in Tennessee), nor did those for CO_2 quadrupling contain more than 0.025 cm (0.01 inch) per month (winter in Ontario). The 100-year cumulative climate changes produced by such low-level annual increments are well beyond any measured in historical times and, indeed, are beyond the measurement capabilities of scientists, who possess a maximum useful lifetime of 40 to 60 years.

Are the forest dynamics shown in Figures 3 to 5 likely to actually occur in the future? The answer depends primarily on the validity of the climate scenario used; assuming that to be correct (a very insecure assumption), then the answer is, "probably not very often." The most important departure from reality is the lack of simulated disturbance. Clearly, certain ecosystem types, such as boreal forests or deciduous sprout-hardwoods, have always undergone irregular, but frequent, catastrophic disturbances by natural forces, such as fire, flood, and windstorms. In addition, extensive, large-scale logging of primeval and secondary forests in about the past 200 years has left a landscape mosaic composed of relatively young, even-aged stands. The earlier simulations indicate that if such young, mature trees dominate forests, their slowed growth response to warming could result in a different and much more limited response of the system to at least initial climate change compared with the responses induced when large numbers of mortality-prone saplings and large, senescent trees dominate the stand structure. Although frequency and intensity of disturbance can be provided in the model with either historical events or random ones (using appropriate statistical distributions of the events), lack of such phenomena in the above simulations significantly reduces their realism.

The model structure itself contains important limitations for the application previously illustrated. For example, the natural history characteristics of tree taxa, documented at the species level, form the basis for all forest stand responses to external environmental variables. Yet, many species behave as groups (ecotypes) of populations, each with slightly different responses to those environmental variables. Although there is a great deal of forestry literature to characterize seed and seedling behavior by "provenance," little information is available on population response differences among mature trees. The model is particularly sensitive to maximum age and diameter data, because these control the annual growth rates and therefore competitive advantages among species at all ages. The lack of intraspecies variation in these two characteristics can produce inaccurate growth responses of the entire stand to the climate changes simulated (for example, behavior of northern white cedar; *see* Model Validity and Figure 4a).

The current model poorly represents the short-term early successional processes that occur before canopy closure. The differences in seedling responses to shading, seed-bed characteristics, soil-surface drought, and temperature from one species to another are crudely represented in the model in terms of two shading categories and of responses by mature trees to the other variables. The limited number of tree species that contribute seeds to development of a real forest stand are replaced in the model by seed sources from all species that could potentially grow in a region. As a result, the model cannot predict, as effectively as can regression techniques, relevant features of short-term, precanopy forest dynamics, productivity, and size structure on specific sites.

Although the specific model presented contains these and other constraints, the general simulation approach holds great potential for future industry analyses and decision making, not only on effects of CO_2 but on acidic deposition and similar questions. The time scales involved in producing a single generation of trees and the increasing anthropocentric problems at regional and continental spatial scales severely reduce the validity of empirical analyses that are based on a few years (or decades!) of environmental measurements at a limited number of locations. Rather, approaches must be developed for predicting the biotic responses by imitating the processes from which forest responses are derived. The above simulations illustrate that approach, and they represent an initial step toward that goal.

Such an approach is particularly relevant to an industry that characteristically (1) is constrained by economic uncertainties to long-term planning covering less than a decade (Woodman, this volume), (2) depends upon the long-term growth characteristics of the forest resource, and (3) consists of many small companies, each having limited resources for mitigating (by silvicultural and genetic research or development of new techniques) future problems. Large corporations can survive impending "surprises" that ruin small businesses. Today, the judgment of which known and uncertain problems (economic, legislative, environmental, or biological) will create the primary threats to the survival of individual small companies is exceedingly complex. One solution to foreseeing the nature and ramifications of those problems, before they develop, is simulation modeling of the entire system (environmental, ecological, and economic). The model approach presented here can provide the necessary information on the ecological component of this system.

ACKNOWLEDGMENTS

We thank M. Lynn Tharp for the model programming and execution, and T. H. Smith and R. J. Norby for constructive reviews. Research was sponsored jointly by the National Science Foundation under interagency agreement no. DEB81-15316 and by the Division of Carbon Dioxide Research, Office of Basic Energy Sciences, U.S. Department of Energy, under contract no. DE-AC05-840R21400 with Martin Marietta Energy Systems, Inc.

REFERENCES

Aber, J. D., D. B. Botkin, and J. M. Melillo. 1978. Predicting the effects of different harvesting regimes on forest floor dynamics in northern hardwoods. *Can. J. For. Res.* 8, 306-315.

Aber, J. D. and J. M. Melillo. 1982. Nitrogen immobilization in decaying hardwood leaf litter as a function of initial nitrogen and lignin content. *Can. J. Bot.* 60, 2263-2269.

Bassett, J. R. 1964. Tree growth as affected by soil moisture availability. *Soil Sci. Soc. Am. Proc.* 28, 436-438.

Bormann, F. H. 1965. Changes in growth pattern of white pine trees undergoing suppression. *Ecology* 46, 269-277.

Botkin, D. B., J. F. Janak, and J. R. Wallis. 1972. Some ecological consequences of a computer model of forest growth. *J. Ecol.* 60,

849-872.

Braun, E. L. 1950. *Deciduous Forests of Eastern North America.* Hafner Press, New York.

Bray, J. R. 1956. Gap-phase replacement in a maple-basswood forest. *Ecology* 37, 598-600.

Camp, W. H. 1951. A biogeographic and paragenetic analysis of American beech (*Fagus*). Am. Phil. Soc. Yearb. 1950, 166-169.

Cohen, S. B., ed. 1973. *Oxford World Atlas.* Oxford University Press, New York.

Copeland, O. L., Jr. 1952. Root mortality of shortleaf and loblolly pine in relation to soils and littleleaf disease. *J. For.* 50, 21-25.

Doyle, T. W. 1983. Competition and growth relationships in a mixed-age, mixed species forest community. Ph.D. dissertation. University of Tennessee, Knoxville.

Fowells, H. A. 1965. *Silvics of Forest Trees of the United States.* Agriculture Handbook no. 271. USDA Forest Service, Washington, D.C.

Fritts, H. C. 1976. *Tree Rings and Climate.* Academic Press, Inc., New York.

Gammon, R. H., W. D. Komhyr, L. S. Waterman, T. Conway, and K. Thoning. 1984. Estimating the natural variation in atmospheric CO_2 since 1860 from interannual changes in tropospheric temperature and the history of major El Niño events. *In* E. T. Sundquist and W. S. Brocker (ed.), *Proceedings, Chapman Conference on Natural Variations in Carbon Dioxide and the Carbon Cycle.* American Geological Union, Washington, D.C.

Hare, K. F., and M. K. Thomas. 1979. *Climate Canada*, 2nd ed. John Wiley and Sons Canada Ltd., Toronto.

Harlow, W. M., E. S. Harrar, and F. M. White. 1979. *Textbook of Dendrology*, 6th ed. McGraw-Hill, New York.

Hepting, G. H. 1963. Climate change and forest diseases. *Annu. Rev. Phytopathol.* 1, 31-50.

Hicks, R. R. 1980. Climate, site and stand factors, p. 55-68. *In* R. C. Thatcher, J. L. Searcy, J. E. Coster, and G. D. Hertel (ed.), *The Southern Pine Beetle.* USDA Forest Service Technical Bulletin no. 1631. USDA, Washington, D.C.

Jacobs, M. R. 1955. Growth habits of eucalyptus. *Aust. Forest. Timber Bur.* 1955, 1-262.

Keeling, C. D., R. B. Bacastow, and T. P. Whorf. 1982. Measurements of the concentration of carbon dioxide at Mauna Loa Observatory, Hawaii, p. 377-385. *In* W. C. Clark (ed.), *Carbon Dioxide Review: 1982.* Oxford University Press, New York.

Kessell, S. R. 1979. Adaptation and dimorphism in eastern hemlock, *Tsuga canadensis* (L.) Carr. *Am. Nat.* 113, 333-350.

Kozlowski, T. T. 1969. Tree physiology and forest pests. *J. For.* 69, 118-122.

Kramer, P. J., and J. P. Decker. 1944. Relation between light intensity and rate of photosynthesis of loblolly pine and certain hardwoods. *Plant Physiol.* 19, 350-358.

Kramer, P. J., and T. T. Kozlowski. 1979. *Physiology of Trees*, p. 109, 381. McGraw-Hill, New York.

Kuchler, A. W. 1964. Potential natural vegetation of the conterminous United States. Spec. Publ. 36. American Geographic Society, New York.

Larcher, W. 1969. The effect of environmental and physiological variables on the carbon dioxide gas exchange of trees. *Photosynthetica* 3, 167-198.

Lindsey, A. A., and J. E. Newman. 1956. Use of official weather data in spring time-temperature analysis of an Indiana phenological record. *Ecology* 38, 812-823.

Little, E. L. 1971. *Atlas of United States Trees, vol. I: Conifers and Important Hardwoods.* USDA Forest Service Misc. Publ. 1146. Washington, D.C.

Little, E. L. 1977. *Atlas of United States Trees, vol. IV: Minor Eastern Hardwoods.* USDA Forest Service Misc. Publ. 1342. Washington, D.C.

Manabe, S., and R. J. Stouffer. 1980. Sensitivity of a global model to an increase of CO_2 in the atmosphere. *J. Geophys. Res.* 85(C10), 5529-5554.

Manabe, S., and R. T. Wetherald. 1980. On the distribution of climate change resulting from an increase in CO_2-content of the atmosphere. *J. Atmos. Sci.* 37, 99-118.

Manabe, S., R. T. Wetherald, and R. J. Stouffer. 1981. Summer dryness due to an increase of atmospheric CO_2-concentration. *Clim. Change* 3, 347-386.

Mattson, W. J. 1980. Herbivory in relation to plant nitrogen content. *Annu. Rev. Ecol. Sys.* 11, 119-161.

Mattson, W. J., and N. D. Addy. 1975. Phytophagous insects as regulators of forest primary production. *Science* 190, 515-522.

Mitchell, J. F. B. 1983. The seasonal response of a general circulation model to changes in CO_2 and sea temperatures. *Q. J. R. Meteorol. Soc.* 109, 113-152.

Mitchell, J. F. B., and G. Lupton. 1984. A $4XCO_2$ integration with prescribed changes in sea surface temperatures. *Prog. Biometeorol.* 3, 353-374.

Monsi, M., Z. Uchijima, and T. Oikawa. 1973. Structure of foliage canopies and photosynthesis. *Annu. Rev. Ecol. Syst.* 4, 301-327.

Nichols, J. O. 1968. Oak mortality in Pennsylvania: a ten-year study. *J. For.* 66, 681-694.

Olson, R. J., C.J. Emerson, and M. K. Nungesser. 1980. Geoecology: a county-level experimental data base for the conterminous United States. ORNL/TM-7351. Oak Ridge National Laboratory, Oak Ridge, Tenn.

Palutikof, J. P., T. M. L. Wigley, and J. M. Lough. 1984. Seasonal climate scenarios for Europe and North America in a high-CO_2, warmer world. U.S. Department of Energy. Washington, D.C.

Pastor, J., and W. M. Post. 1984. Development of a linked forest productivity-soil carbon and nitrogen model. Oak Ridge National Laboratory Technical Manuscript, Oak Ridge, Tenn.

Perry, A. M. 1982. Carbon dioxide production scenarios. pp. 335-363. *In* W. C. Clark (ed.), *Carbon Dioxide Review: 1982.* Oxford University Press, New York.

Phipps, R. L. 1979. Simulation of wetlands forest vegetation dynamics. *Ecol. Mod.* 7, 257-288.

Redmond, D. R. 1959. Mortality of rootlets in balsam fir defoliated by spruce budworm. *For. Sci.* 5, 64-69.

Rowe, J. S. 1972. *Forest Regions of Canada.* Publication 1300. Canadian Forestry Service, Ottawa.

Runkle, J. R. 1981. Gap regeneration in some old-growth forests of the eastern United States. *Ecology* 62, 1041-1051.

Sakai, A., and C. J. Weiser. 1973. Freezing resistance of trees in North America with reference to tree regions. *Ecology* 54, 118-126.

Schmidt, R. A. 1978. Disease in forest ecosystems: the importance of functional diversity, p. 278-315. *In* J. G. Horsfall and E. B. Cowling (ed.), *Plant Disease: an Advanced Treatise, vol. II.* Academic Press, Inc., New York.

Schneider, S. H., and S. L. Thompson. 1981. Atmospheric CO_2 and climate: importance of the transient response. *J. Geophys. Res.* 8(C4), 3135-3147.

Schowalter, T. D., R. N. Coulson, and D. A. Crossley. 1981. The role of southern pine beetle and fire in maintenance of structure and function of the southeastern coniferous forest. *Environ. Entomol.* 10, 821-825.

Shugart, H. H. 1984. *A Theory of Forest Dynamics.* Springer-Verlag, New York.

Shugart, H. H., and D. C. West. 1977. Development of an Appalachian deciduous forest succession model and its application to assessment of the impact of the chestnut blight. *J. Environ. Manage.* 5, 161-179.

Shugart, H. H., and D. C. West. 1979. Size and pattern of simulated forest stands. *For. Sci.* 25, 120-122.

Sollins, P., D. E. Reichle, and J. S. Olson. 1973. Organic matter budget and model for a southern Appalachian *Liriodendron* forest. EDEB/IBP-73/2. Oak Ridge National Laboratory, Oak Ridge, Tenn.

Solomon, A. M. 1982. Plant community response to decreased seasonality during full-glacial time. Abstr. no. 18. 7th Bienn. Meeting, Amer. Quat. Assoc., Seattle, Wash.

Solomon, A. M., H. R. Delcourt, D. C. West, and T. J. Blasing. 1980. Testing a simulation model for reconstruction of prehistoric forest stand dynamics. *Qual. Res.* 14, 275-293.

Solomon, A. M., D. C. West, and J. A. Solomon. 1981. Simulating the role of climate change and species immigration in forest succession, p. 154-177. *In* D. C. West, H. H. Shugart, and D. B. Botkin (ed.), *Forest Succession: Concepts and Application.* Springer-Verlag, New York.

Solomon, A. M., and H. H. Shugart. 1984. Integrating forest-stand simulations with paleoecological records to examine long-term forest dynamics, p. 333-356. *In* G. I. Agren (ed.), *State and Change of Forest Ecosystems*. Swedish University of Agricultural Science, Ecological and Environmental Research Report no. 13. Uppsala, Sweden.

Solomon, A. M., and M. L. Tharp. 1984. Carbon storage response of forest communities to mid-latitude climate changes of the past 20,000 years: simulation model results. *In* E. T. Sundquist and W. S. Brocker (ed.), *Proceedings, Chapman Conference on Natural Variations in Carbon Dioxide and the Carbon Cycle.* American Geological Union, Washington, D.C. (in press).

Solomon, A. M., M. L. Tharp, D. C. West, G. E. Taylor, J. M. Webb, and J. C. Trimble. 1984. Response of unmanaged forests to CO_2 induced climate change: available information, initial tests, and data requirements. U.S. DOE TR009, NBB-0053. U.S. Department of Energy, Washington, D.C.

Tharp, M. L. 1978. Modeling major perturbations in a forest ecosystem. M.S. thesis. University of Tennessee, Knoxville.

Thornthwaite, C. W., and J. R. Mather. 1957. Instructions and tables for computing potential evapotranspiration and the water balance. *Publ. Climatol.* 10, 183-311.

Tyrone, E. H. and R. P. True. 1968. Radial increment response of Appalachian hardwood species to a spring freeze. *J. For.* 66, 488-591.

U.S. Department of Agriculture. 1982. An analysis of the timber situation in the United States, 1952-2030. Forest Service Resource Report no. 23. U.S. Department of Agriculture, Washington, D.C.

Watt, A. S. 1947. Pattern and process in the plant community. *J. Ecol.* 35, 1-22.

Weinstein, D. A., H. H. Shugart, and D. C. West. 1982. The long-term nutrient retention properties of forest ecosystems: simulation investigation. ORNL/TM-8472. Oak Ridge National Laboratory, Oak Ridge, Tenn.

White, T. C. R. 1969. An index to measure weather-induced stress of trees associated wth outbreaks of psyllids in Australia. *Ecology* 50, 905-909.

White, T. C. R. 1974. A hypothesis to explain outbreaks of looper caterpillars, with special reference to populations of *Selidosema suavis* in a plantation of *Pinus radiata* in New Zealand. *Oecologia* 16, 279-301.

Wigley, T. M. L. 1983. The pre-industrial carbon dioxide level. *Clim. Change* 5, 315-320.

Wigley, T. M. L., P. D. Jones, and P. M. Kelly. 1980. Scenario for a warm, high-CO_2 world. *Nature* 283, 17-21.

Zahner, R., and J. R. Donnelly. 1967. Refining correlations of water deficits and radial growth in young red pine. *Ecology* 48, 525-530.

EFFECTS OF INCREASING CARBON DIOXIDE CONCENTRATION ON THE PHYSIOLOGY AND GROWTH OF FOREST TREES

by Paul J. Kramer and Nasser Sionit

ABSTRACT

Short-term experiments indicate that increasing the concentration of atmospheric carbon dioxide (CO_2) increases stem diameter, height, and dry weight of seedlings of several tree species. A high CO_2 concentration also increases branching, leaf area, and leaf thickness in some but not all species. Root-shoot ratio does not seem to be significantly affected. At the end of the first growing season, dry weights of roots, stems, and leaves of sweetgum and loblolly pine seedlings were greater at 500 than at 350 parts per million (ppm) of CO_2. Sweetgum also produced more branches and leaves at the higher concentration. There was no significant increase in dry weight of roots, stems, or leaves at 650 ppm over the value at 500 ppm, but there was a significant increase in height of loblolly pine seedlings. It seems probable that if other environmental factors are not severely limiting, the increasing concentration of atmospheric CO_2 will significantly increase tree growth, but the magnitude of the effects will vary among species and with the severity of other environmental stresses. Although increasing concentration of CO_2 may partly compensate for low light, high temperature, and water stress and possibly even for mineral deficiencies, unfavorable factors such as droughts, abnormal temperatures, and unfavorable soil conditions will continue to limit tree growth. There is need for more long-term research to provide reliable information concerning the effects of CO_2 buildup on the relative growth of commercially important species.

Foresters have learned from experience how to cope reasonably well with environmental stresses, such as drought, waterlogged soil, mineral deficiencies, and unfavorable temperatures, by choosing appropriate species and management practices. Now a new environmental factor, the increasing concentration of atmospheric carbon dioxide (CO_2), has been identified. It is present worldwide and is inexorably increasing in importance. Unfortunately, the CO_2 problem has been identified only recently, and it is too new to have been evaluated fully. There is a tendency to postpone consideration of its effects until the future, when the concentration will have doubled, but trees are being affected now. The CO_2 concentration has increased about 15% during the past 50 years and probably will rise more rapidly during the next 50 years. The effects of this increase may already have been detected in the annual rings of subalpine conifers (LaMarche et al., 1984).

Presumably the major question for foresters is whether the changes in relative growth rates and wood production among the important commercial species will be sufficiently important to justify changes in choice of species for future planting. They need to know whether the tree species most successful today will be those most successful 50 or 100 years from now when the CO_2 concentration may exceed 500 parts per million (ppm). Various secondary questions concern effects on drought tolerance, mineral nutrient requirements, wood quality, branching, and damage from pests, all of which might affect management plans.

KNOWN EFFECTS OF HIGH CO_2 CONCENTRATION ON GROWTH

Greenhouse crops such as lettuce and tomatoes have been "fertilized" successfully with CO_2 for many years, and much research has been done on herbaceous plants. The early work was summarized by Miller (1938). Kimball (1983) surveyed the recent literature and collected 70 reports and over 340 observations on the effects of enhanced CO_2 on 24 agricultural crops and 14 other species. Evaluation of these data led him to conclude that the marketable yield of these herbaceous plants would be increased an average of 33% by doubling the CO_2 concentration of the atmosphere. However, this estimate seems to assume favorable environmental conditions. In contrast, Waggoner (1984) expects very small net effects on crop yields because of the compensatory effects of climatic changes and environmental stresses.

Unfortunately, the data for woody plants are much more limited than those for herbaceous plants and are based largely on seedling experiments that lasted only for a few weeks or months. While seedling quality and establishment are important, experiments lasting, at most, one growing season are too short to permit safe generalizations concerning what will occur later in development. Sionit and Kramer (1984) summarized the results of 15 of these short-term experiments on woody plants and reported the following general conclusions with respect to growth.

- Dry weight, stem diameter, and height of seedlings generally are increased by CO_2 concentrations in the range of 400 to

700 ppm.

- CO_2 concentrations above 500 to 600 ppm usually produce little additional growth and sometimes even cause a decrease.
- Seedlings of various species react differently to increases in CO_2 concentration, making it necessary to study each species individually.
- A high concentration of CO_2 seems to have morphological effects, causing increasing branching, leaf area, and leaf thickness in some but not all species.
- The competitive capacity of seedlings of various species is likely to be affected differently by an increase in CO_2 concentration.

Fortunately, there seem to be some similarities in the reaction to high CO_2 between C_3 herbaceous and woody plants. This was shown in a study at North Carolina State University by Rogers and his associates (1983) in which corn (*Zea mays* L.), soybeans (*Glycine max* L. Merr.), and sweetgum (*Liquidambar styraciflua* L.) were grown outdoors in open-topped containers and maintained at various CO_2 concentrations for 12 or 13 weeks. They found that stomatal conductance decreased about 50% in all three species as the CO_2 concentration increased from ambient to 910 ppm. At the same time the rate of net photosynthesis increased 65% for sweetgum and about 70% for soybean.

The failure of corn to show any increase relates to its C_4 carbon-fixation pathway. Corn, sugarcane, and a number of other plants have what is called the Hatch-Slack or C_4 carbon pathway because CO_2 is fixed initially into four-carbon organic acids in the leaf mesophyll cells by an enzyme, phosphoenolpyruvate carboxylase (PEP carboxylase), which is saturated at the present atmospheric concentration of CO_2 and is little affected by further increases. Also, in C_4 plants the organic acid produced by carboxylation in the mesophyll cells diffuses to the bundle sheath cells and is decarboxylated, producing a high concentration of CO_2 that increases the rate of the final conversion to sugars by the Calvin cycle or C_3 pathway. This results in a very low apparent mesophyll resistance for CO_2 in C_4 plants, making them less responsive than C_3 plants to an increase in atmospheric CO_2 concentration.

However, the important point here is that the behavior of the two C_3 plants, herbaceous soybean and woody sweetgum, is surprisingly similar. This encourages extrapolation from herbaceous plants to woody seedlings until more data become available for the latter. For example, in the experiments of Rogers et al. (1983), the dry weight of both loblolly pine and sweetgum increased about 40%

at 910 ppm. In other experiments by Tolley and Strain (1984a) and Sionit et al. (1985), dry weights of both species also increased significantly at CO_2 concentrations of 500 to 1000 ppm (Table 1).

Long-term experiments are now in progress in the Duke University Phytotron. Seedlings of loblolly pine (*Pinus taeda* L.) and sweetgum were germinated and grown in controlled-environment greenhouses at 350, 500, and 650 ppm of CO_2, with temperature regimes following those outside and with natural photoperiods and light regimes. At the end of the first growing season the dry weights of sweetgum seedlings were over 40% greater and those of loblolly pine were nearly 60% greater at 500 than at 350 ppm of CO_2 (Table 1). However, no significant increase in dry weights occurred at 650 ppm of CO_2 over the weights at 500 ppm. Root-shoot ratios were not significantly increased by CO_2 enrichment in either species. Stem height increased with increasing CO_2 in both species, and sweetgum was taller than loblolly pine at all concentrations (Figure 1). Sweetgum seedlings produced significantly more leaves at the high CO_2 concentrations than at 350 ppm (Figure 2), probably because they produced more branches. Leaf senescence began about the same time in 350 and 650 ppm of CO_2.

The data in Table 2 show that all the organs of loblolly pine seedlings continued to increase in dry weight during the winter. The rate of net photosynthesis of pine needles was greater at 650 ppm of CO_2 than at lower CO_2 concentrations, but the percent increase in dry weight was no greater at 650 than at 350 ppm. Sweetgum seedlings showed no important changes in dry weight over winter if the total dry weight was corrected for leaf loss (Table 3). Data from another winter are needed to establish whether the small loss in dry weight at 350 ppm and the gain at 650 ppm are real.

GROWTH EFFECTS OF HIGH CO_2 CONCENTRATIONS

The data available suggest that doubling the atmospheric CO_2 concentration will increase the growth of tree seedlings of at least some species as much as it increases the growth of herbaceous plants, if other environmental factors are not severely limiting. It is a general principle that the quantity and quality of growth are controlled by the interaction of the hereditary potential (the genome) and the environment, operating through basic physiological processes. Environmental stresses, such as drought or unfavorable temperature, and silvicultural treatments, such as thinning, pruning, and fertilization, affect growth by modifying the physiological processes and conditions that directly control growth

Figure 1

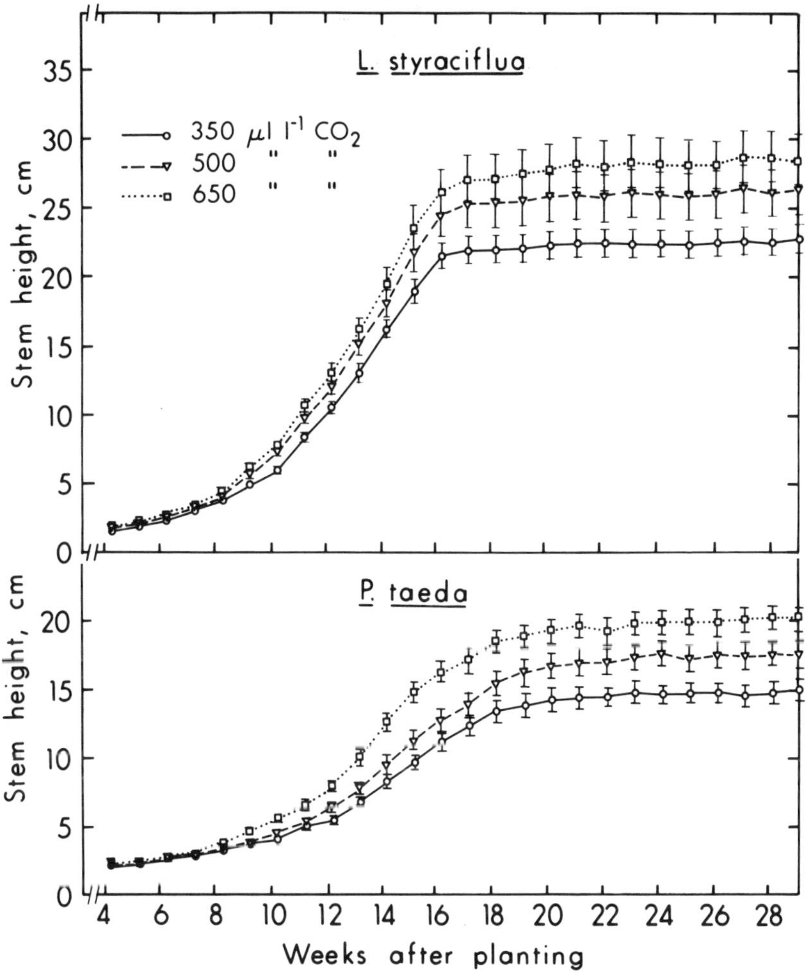

Effects of increasing concentrations of CO_2 on height of sweetgum and loblolly pine grown from seed in an air-conditioned greenhouse with natural photoperiod and temperatures simulating the two-week average obtained from a 30-year temperature record. The seedlings became dormant and the sweetgum shed their leaves by about week 32. Concentration of CO_2 as microliters per liter ($\mu l\ l^{-1}$) is equivalent to ppm. Figures 1 and 2 are based on measurements of 15 seedlings at each date. (From data of Sionit et al., 1985.)

Figure 2

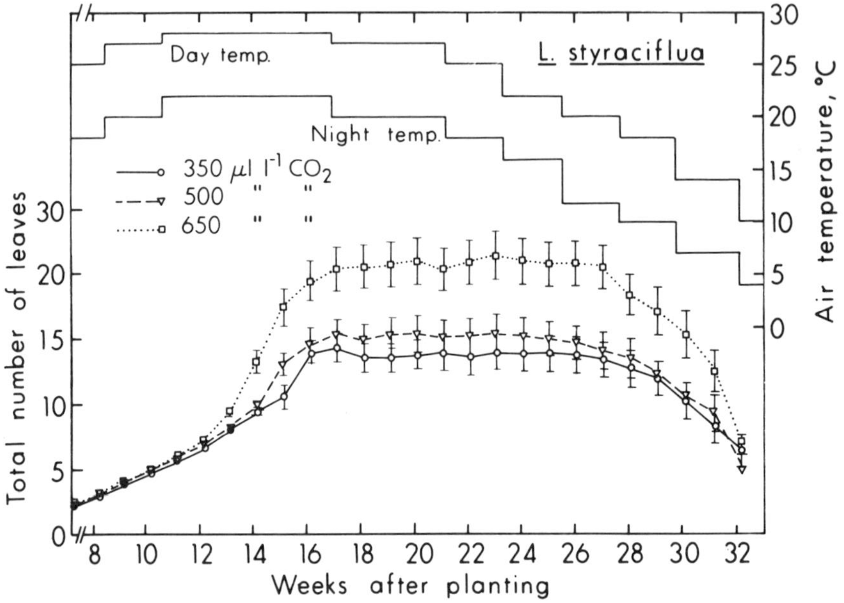

Temperature regime and effects of increased concentration of CO_2 on numbers of leaves on sweetgum seedlings in the experiment described under Figure 1. The increase in leaf number at 650 $\mu l\ l^{-1}$ (ppm) was correlated with the increase in number of branches. (From data of Sionit et al., 1985.)

Table 1
Effects of Atmospheric Carbon Dioxide Enrichment on Growth of *L. Styraciflua* (Sweetgum) and *P. Taeda* (Loblolly Pine) at 32 Weeks after Planting (end of first growing season)

Species	Carbon dioxide concentration (ppm)	Stem weight (g)	Leaf weight (g)	Root weight (g)	Root-shoot ratio	Total weight (g)	Percent increase in total weight
L. styraciflua	350	2.1 a	1.5a	6.3 a	1.7 ab	9.9 a	
	500	3.4 b	1.6 ab	9.2 b	1.8 b	14.2 b	43.4
	650	3.7 b	2.0 b	8.6 ab	1.5 a	14.3 b	44.4
P. taeda	350	0.4 a	1.4a	2.1 a	1.2 a	3.9 a	
	500	1.0 b	2.3 b	2.9 b	0.9 a	6.2 b	58.9
	650	0.8 b	2.1 b	3.2 b	1.1 a	6.1 b	56.4

Means followed by the same letter within a column for each species are not significantly different at the 5% level of probability. Values are means of measurements on 15 seedlings. From Sionit et al., 1985.

Table 2
Effects of Atmospheric Carbon Dioxide Enrichment on Growth of *L. Styraciflua* (Sweetgum) and *P. Taeda* (Loblolly Pine) at 48 Weeks after Planting (end of first dormancy period)

Species	Carbon dioxide concentration (ppm)	Stem weight (g)	Leaf weight (g)	Root weight (g)	Root/shoot ratio	Total weight (g)
L. styraciflua	350	1.8 a	--	5.9 a	3.3 b	7.7 a
	500	3.0 b	--	9.2 b	3.1 ab	12.3 b
	650	3.5 b	--	9.5 b	2.8 a	13.0 b
P. taeda	350	1.0 a	2.2 a	4.1 a	1.3 a	7.3 a
	500	1.2 a	2.5 a	4.8 ab	1.3 a	8.5 ab
	650	1.5 a	3.3 b	6.5 b	1.4 a	11.3 b

Means followed by the same letter within a column for each species are not significantly different at the 5% level of probability. From Sionit et al., 1985.

Table 3
Changes over the Winter Period in Total Dry Weight of L. Styraciflua (Sweetgum) and P. Taeda (Loblolly Pine) Seedlings Grown at Three Atmospheric Carbon Dioxide Concentrations[1]

Species	Carbon dioxide concentration (ppm)	Percent change in total dry weight	Net photosynthesis (mg g-1 DW h-1)[2]
L. styraciflua[3]	350	-8.4	-
	500	-2.4	-
	650	+5.7	-
P. taeda	350	+87.2	1.4
	500	+37.1	1.9
	650	+85.2	3.4

[1]From data of Sionit et al., 1984.

[2]Milligrams per gram of dry weight per hour.

[3]Corrected for leaf loss.

(Figure 3).

Carbon dioxide is no exception to this general principle, so its effects on growth must be explained in terms of its effects on important physiological conditions and processes including stomatal conductance, photosynthesis, carbohydrate partitioning, and the plant's water economy. An increase in CO_2 concentration directly affects stomatal conductance and photosynthesis and indirectly affects transpiration and other processes discussed later.

Direct Effects

Experimental evidence indicates the following direct effects of CO_2 buildup in tree physiology.

Stomatal Conductance

In most experiments, increasing the CO_2 concentration causes partial closure of stomata, decreasing the conductance for CO_2 and water vapor. In one outdoor experiment the decrease amounted to 40% for sweetgum when the CO_2 concentration was doubled (Rogers et al., 1983). In a growth chamber experiment stomatal conductance was decreased about 50% in sweetgum, but not at all in loblolly pine, by increasing the CO_2 concentration from 350 to 675 ppm (Tolley, 1982). Beadle et al. (1979) reported that the stomatal conductance of Sitka spruce (*Picea sitchensis*) was unaffected by CO_2 concentrations in the range of 20 to 600 ppm. The reasons for differences among species in stomatal response are unknown, probably because the exact mechanism by which high ambient concentrations of CO_2 decrease stomatal aperture is not fully understood.

The complex relationships among stomatal conductance, rate of CO_2 uptake, and transpiration are shown in Figure 4 for Eastern cottonwood (*Populus deltoides*) seedlings subjected to various concentrations of CO_2 for two weeks. Stomatal conductance decreased about 25% between 300 and 600 ppm; transpiration also decreased about 25% but photosynthesis increased about 40%.

Both inward diffusion of CO_2 and outward diffusion of water vapor are affected by stomatal aperture or conductance. Thus it seems paradoxical that photosynthesis can increase when stomatal conductance and transpiration are decreasing. However, the paradox can be explained in part by the differences in resistances to water and CO_2 movement, shown in Table 4, and in part by the increase in CO_2 concentration within leaves. The major resistance to the escape of water vapor from leaves is the stomata, and an

Figure 3

How Environmental Factors Affect Tree Growth

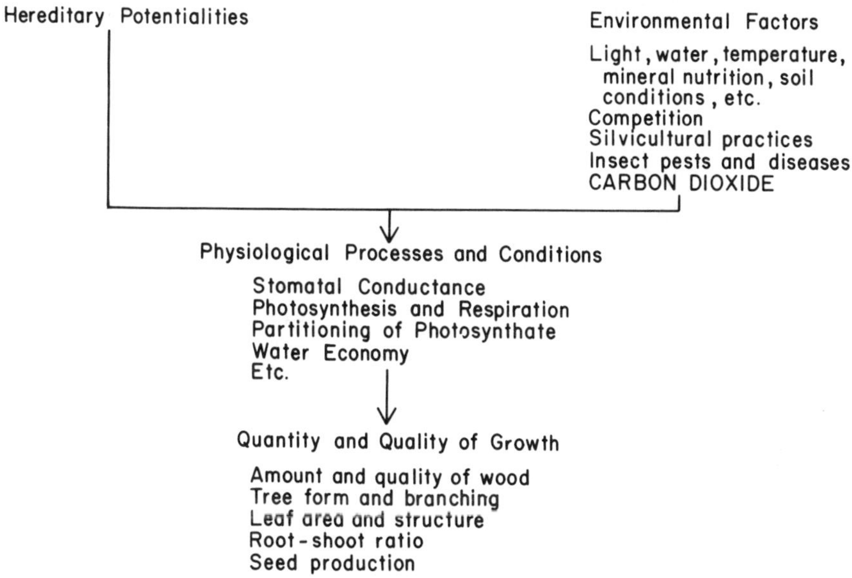

Diagram showing how environmental factors interact with hereditary potentialities to modify the physiological processes and conditions that control the quantity and quality of growth.

Figure 4

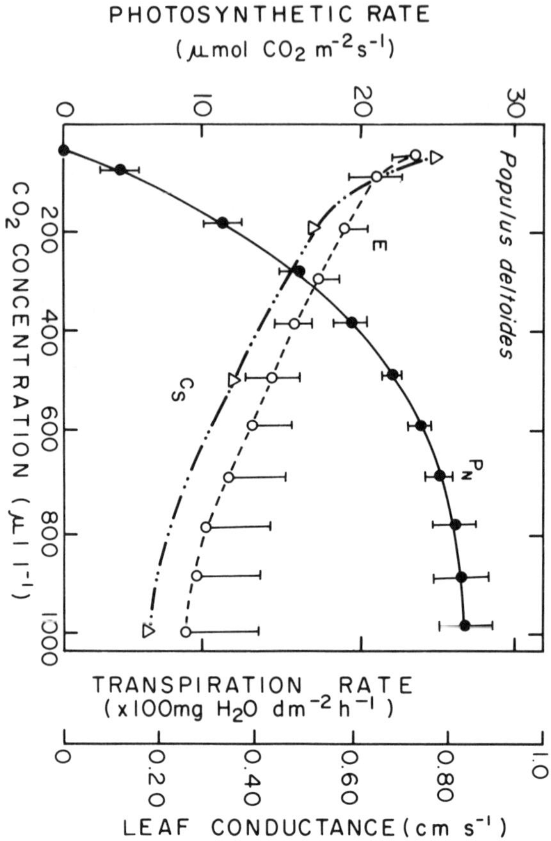

Photosynthesis (P_n), transpiration (E), and leaf conductance (C_s) as functions of CO_2 concentration. (Adapted from Regehr et al., 1975.)

Table 4
Effects of Partial Closure of Stomata on Rates of Transpiration and Photosynthesis

	Air and stomatal resistance (sec cm-1)[2]		Internal or mesophyll resistance (sec cm-1)		Total resistance (sec cm-1)	Reduction in conductance (percent)
H_2O						
Stomata open	2.3	+	0	+	2.3	50
Stomata partly closed	4.6	+	0	+	4.6	
CO_2						
Stomata open	3.7	+	6.1	+	9.8	27
Stomata partly closed	7.4	+	6.1	+	13.5	

[1]From data of Cooke and Rand (1980).

[2]Seconds per centimeter.

increase in stomatal resistance decreases transpiration proportionately. However, there are additional resistances to the movement of CO_2 to the chloroplasts and its fixation into carbohydrates. These are known collectively as the apparent mesophyll resistance, and it is relatively high in trees and in those crop plants possessing the C_3 photosynthetic pathway. Thus a 50% increase in stomatal resistance decreases transpiration about 50%, but decreases photosynthesis much less because it is only part of the total resistance to CO_2 movement. Also, for reasons discussed in the next section, the increase in stomatal resistance is compensated in C_3 plants by the increased internal CO_2 concentration at high external concentrations. This difference in effect of partial closure of stomata on diffusion of water vapor and CO_2 also is the basis for the use of antitranspirants that cause stomatal closure (Kramer, 1983).

Photosynthesis

Most crop plants have the C_3 or Calvin cycle pathway of photosynthesis, in which CO_2 is fixed in three-carbon phosphoglyceric acid by the enzyme ribulose bisphosphate carboxylase (RuBP carboxylase). This enzyme has an affinity for both CO_2 and O_2 and functions both as a carboxylase and as an oxygenase. The oxygenase activity is responsible for photorespiration, which according to various authors can use 15 to 50% of the carbon fixed by photosynthesis in C_3 plants (Gifford et al., 1984). However, an increase in ambient CO_2 concentration results in an increase in CO_2 concentration within leaves, increasing the rate of carboxylation relative to oxygenase activity, which leads to increased net photosynthesis.

Increase in ambient CO_2 concentration also tends to compensate for the inhibitory effects of shading (Tolley and Strain, 1984a) and high temperature. Although the maximum absolute enhancement of photosynthesis by higher concentrations of CO_2 is at saturating light intensities, the maximum relative increase occurs at low intensities (Pearcy and Bjorkman, 1983), where the increased rate of photosynthesis resulting from a higher CO_2 concentration at the carboxylating sites is most important relatively. Thus the increase in atmospheric CO_2 concentration may be most important for photosynthesis in the lower part of tree canopies and for vegetation on the forest floor. Also the optimum temperature for photosynthesis of at least some C_3 plants is increased significantly by an increase in concentration of CO_2 because of the speedup in carboxylation rate and slowdown in photorespiration. This may become more important if increase in the atmospheric CO_2 concentration causes a significant rise in temperature.

Short- versus Long-Term Effects on Photosynthesis. Short-term measurements of photosynthesis, such as those shown in Figure 4, are not always good indicators of the effects of prolonged exposure. Kramer (1981) cited instances in which the rate of photosynthesis of herbaceous species exposed to high concentrations of CO_2 decreased over time; this was observed in loblolly pine and sweetgum seedlings by Tolley and Strain (1984a). In some, but not all, experiments the reduction in rate per unit of leaf area was compensated by increase in leaf area and/or leaf thickness (Tolley and Strain, 1984a; Sionit et al., 1985). One reason for the reduction in rate at high CO_2 concentrations may be feedback inhibition caused by accumulation of carbohydrate (Clough et al., 1981). Another possibility is actual injury to chloroplasts by accumulation of starch, causing distortion of membranes and inhibition of electron transport. This has been observed in several herbaceous species (Hofstra and Hesketh, 1975; Cave et al., 1981; Wulff and Strain, 1982). It also is possible that a high CO_2 concentration reduces the activity or amount of carboxylating enzyme in some plants (Wong, 1979). Thus it is unwise to attempt to extrapolate from short-term experiments to long-term effects of exposure to high concentrations of CO_2.

Photosynthesis and Biomass. The relationship between photosynthesis and biomass at high concentrations of CO_2 was discussed by Kramer (1981). Because biomass depends on photosynthesis, there is a tendency to assume that the chief effect of increasing CO_2 concentration is through increase in the rate of photosynthesis, but the indirect effects may also be important. Biomass is not necessarily closely correlated with rate of photosynthesis because it depends not only on rate per unit of leaf area but also on total leaf area and leaf duration as well as the length of time during which leaves carry on photosynthesis. The latter may become particularly important with respect to competition between evergreen and deciduous trees in regions with mild winters.

It has been pointed out (Figure 4) that the rate of photosynthesis usually increases curvilinearly at intermediate CO_2 concentrations and tends to level off or even decrease at much higher concentrations. However, in some experiments this was compensated for by an increase in leaf area and/or leaf thickness. Sweetgum seedlings grown at high CO_2 concentrations showed some increase in leaf area after 13 weeks, compared to those at 350 ppm, although not after 32 weeks. However, the seedlings showed a significant increase in weight per unit of leaf area at 32 weeks, and this probably indicated an increase in amount of photosynthetic tissue per unit of leaf surface (Nobel, 1980). Loblolly pine also produced a greater weight of needles at the higher concentration of CO_2. Some of these data are shown in Table 1.

Indirect Effects

Increasing the concentration of CO_2 in the atmosphere has various indirect effects not yet fully evaluated.

Partitioning of Carbohydrates

The ultimate success of both field and forest crops depends on partitioning the maximum amount of photosynthate into a merchantable product. In forest trees this usually is bole wood, and the increase in branching observed in sweetgum seedlings probably would be undesirable. The increases in branching, leaf area, and leaf thickness in seedlings grown in high concentrations of CO_2 are caused by changes in the amount of carbohydrate partitioned into various structures, but the causes of these changes are not known.

Although an increase in root-shoot ratio might be expected to accompany an increase in photosynthesis, and has been observed in some herbaceous plants (Sionit et al., 1982, for example), it has not been observed in tree seedlings. No significant differences in root-shoot ratios at various concentrations of CO_2 developed in either pine or sweetgum seedlings during an entire growing season in the experiments of Sionit et al. (1985), for which data are given in Table 1. However, the ratio of roots to shoots was much greater in sweetgum than in loblolly pine at all concentrations of CO_2 (Table 1). Also the root-shoot ratio of loblolly pine seedlings increased substantially over winter at all concentrations of CO_2. The number of branches doubled at 650 ppm on sweetgum seedlings but increased only slightly on loblolly pine seedlings. Although the increase in number of branches on sweetgum seedlings probably represents an undesirable diversion of photosynthate from stem growth, the stem diameter of sweetgum seedlings was significantly greater at the higher concentrations of CO_2. The proportion of the total dry weight occurring in leaves was not consistently affected by increasing concentrations of CO_2, although leaf weight was increased in both species.

In general the overall dry weight of all plant parts increased without any important change in size of any one organ relative to that of another. Perhaps this demonstrates that the correlation in relative size that exists among various organs of a plant is so strong that only a severe stress, such as severe water stress, heavy shading, or serious lack of nitrogen, is likely to cause important changes in proportions of roots to shoots or of leaves versus roots and stems. In some crop plants, the amount of photosynthate partitioned into seeds is important, but there is no information concerning the effects of increasing the CO_2 concentration on flowering and seed production of forest trees. If it causes earlier flowering or a significant increase in seed production, it might be useful in tree-breeding experiments.

Water Relations

Increasing the concentration of CO_2 causes partial closure of stomata and reduces the rate of transpiration. This should postpone dehydration and maintain adequate levels of leaf water potential and cell turgor longer during drought in high than in low concentrations of CO_2. In some short-term experiments this has occurred. In one series of experiments, when seedlings of sweetgum and loblolly pine were water stressed in growth chambers, the water potential was higher in seedlings growing at 1000 ppm than in those growing at 350 ppm, at both high and low irradiances (Tolley, 1982). However, in an unpublished experiment, Sionit found that, although transpiration and stomatal conductance of sweetgum seedlings were reduced at 650 ppm compared with 350 ppm of CO_2, the leaf water potential was decreased to the same extent by water stress at both concentrations of CO_2. Also, no significant differences in water potential were found in loblolly pine seedlings grown at normal and high concentrations of CO_2 and then subjected to water stress. Thus the exact response to water stress in high concentrations of CO_2 seems to depend on presently unidentified factors.

It should be emphasized that stomatal closure merely postpones dehydration. However, there is a possibility that the increased carbohydrate content of leaves in a high concentration of CO_2 may result in increased tolerance of dehydration by causing osmotic adjustment. This means that the osmotic potential decreases more than can be explained by the decrease in water content. The decrease in osmotic potential enables the cells to remain turgid and the stomata to remain at least partially open to a lower water potential. Sionit et al. (1981) found that osmotic adjustment occurs in wheat, but it has not yet been sought for in woody plants subjected to water stress in a high concentration of CO_2. So far as is known, a high concentration of CO_2 confers no increased protoplasmic toleration of dehydration.

Water-Use Efficiency

"Water-use efficiency" (WUE) is a term widely used today in discussions of plant-water economy (*see* Sinclair et al., 1984). As used by physiologists, the term refers to units of CO_2 fixed per unit of water transpired. On this basis, an increase in CO_2 concentration is likely to materially increase WUE because it tends to decrease transpiration and increase photosynthesis. Downton et al. (1980) reported that doubling the CO_2 concentration doubled the WUE of creosote bush (*Larrea divaricata*) and oleander (*Nerium oleander*) and that the differences were maintained for two months. Rogers et al. (1983) reported that the WUE of sweetgum, corn, and soybeans growing in open-topped containers in the field was more than doubled at 910 ppm, and Tolley (1982) observed that the WUE of sweetgum seedlings was more than doubled at 1000 ppm of CO_2 in growth chambers.

Thus, in terms of the ratio of leaf photosynthesis to leaf transpiration, there appears to be a dramatic increase in WUE. However, Sinclair et al. (1984) discussed the difficulties in extrapolating from leaves to leaf canopies, and it is uncertain whether conclusions based on essentially instantaneous measurements made on single leaves can be extrapolated to forest canopies without considerable qualifications. It seems possible that leaves at lower levels in a canopy may have a WUE quite different from fully exposed leaves. Some of the uncertainties encountered in estimating transpiration within a forest canopy are discussed by Kaufmann (1984), and similar uncertainties exist with respect to photosynthesis. Furthermore, if there is an increase in leaf area, this may partially offset the decreasing rate of transpiration. Also, for practical purposes the WUE of a stand depends on the relationship between wood produced and total water used in evaporation and transpiration. The total water used often is expressed for watersheds as the difference between precipitation and runoff. How good the correlation is between WUE of a forest stand and that based on physiological measurements of a single leaf remains to be determined.

INTERACTION OF CO_2 CONCENTRATION WITH OTHER LIMITING FACTORS

In the forest, tree growth usually is limited by a combination of stresses. At present the suboptimal concentration of atmospheric

CO_2 always is limiting, but, in addition, growth often is limited by drought, mineral deficiencies, shade, and unfavorable temperatures, either singly or in combination. The concept of limiting factors sometimes is incorrectly invoked to argue that if growth already is limited by one suboptimal factor it cannot respond to an increase in another suboptimal factor, such as CO_2 concentration. This is not true unless the other stresses are very severe. According to Mitscherlich's concept of limiting factors, if growth is limited simultaneously by several factors, the response to an improvement in any one factor will be approximately proportional to the degree of deficiency of that factor.

An increase in CO_2 concentration will produce a measurable increase in growth of both woody and herbaceous plants suffering from water, temperature, or mineral deficiency stress. For example, Tolley and Strain (1984b) found that even with limiting irradiance (250 μmol m^{-2} s^{-1}) and water stress, doubling the ambient CO_2 concentration produced a 17% increase in dry weight of sweetgum seedlings and a small increase in pine seedlings, as shown in Table 5. However, in the absence of water or light stress, doubling the ambient CO_2 concentration produced more than twice as much increase in dry weight as for stressed plants. Thus, while an increase in CO_2 concentration increased growth of water- and light-stressed seedlings, it produced a greater increase in unstressed seedlings. Other examples of such interactions will be discussed next.

Water Stress

Probably water stress is the factor most often limiting tree growth, both directly by stomatal and nonstomatal inhibition of photosynthesis and indirectly by reducing cell expansion and decreasing leaf area and shoot and root elongation. In general, cell enlargement and growth are inhibited by water stress before photosynthesis is much reduced (Boyer, 1970). The reduced sink capacity caused by reduced growth is likely to result in accumulation of carbohydrate and reduction in photosynthesis by feedback inhibition, regardless of CO_2 concentration. Also, because roots are subjected to less water stress than shoots, roots are likely to make more growth in mildly water-stressed plants, resulting in an increased root-shoot ratio (Sharp and Davies, 1979). If water stress reduces leaf expansion early in the growing season, this may reduce the benefit expected from increased CO_2 concentration. Thus, although in theory doubling the CO_2 concentration should reduce injury from drought, it is uncertain how much it

Table 5
Compensatory Effects of Increased Carbon Dioxide Concentration on Dry Weight of Stressed Sweetgum and Loblolly Seedlings[1]

Species	Carbon dioxide (ppm)	Light (μmol $m^{-2}s^{-1}$)[2]	Water potential (bars)	Total dry weight (g)	Percent increase
L. styraciflua	350	250	-22	0.81	
	675	250	-10	0.95	17
	675	1000	-22	1.06	30
	675	1000	-3	1.15	42
P. taeda	350	250	-18	0.36	
	675	250	-18	0.39	8
	675	1000	-24	0.51	41
	675	1000	-7	0.62	72

[1]n = 10 (Tolley and Strain, 1984b).

[2]Micromoles per square meter per second.

will compensate for the effects of the severe droughts characteristic of much of the United States.

The only data available on the interaction of water stress and CO_2 on tree seedlings are those in Table 5, which indicate that doubling the CO_2 concentration compensated to some degree for water and light stress either singly or in combination. The data also suggest that low irradiance is more limiting to loblolly pine that to sweetgum.

Temperature

The compensating effects of increased CO_2 for high temperatures were discussed briefly in relation to photosynthesis and are further discussed by Leverenz and Lev (this volume). However, a few special examples should be mentioned. Kriedemann et al. (1976) reported that placing grapes and several other kinds of woody plants in a CO_2 concentration of 1200 to 1300 ppm increased their growth and tolerance of temperatures in the range of 37° to 40°C (98.5° to 104°F). Only guava, among a half dozen plants tested, showed no improvement in growth. The grapes were exposed to these abnormally high temperatures for 3 to 6 months to eliminate viruses. Sionit et al. (1981) reported that okra (*Abelmoschus esculentus* L. Moench) is unproductive when grown at temperatures lower than 26/20°C (79/68°F) (day/night) in ambient CO_2, but in a concentration of 1000 ppm of CO_2 it yielded well at 20/14°C (68/57°F) (day/night). Thus high concentrations of CO_2 seem to have considerable compensatory effects for both high and low temperatures, although the effects probably will be greater at high than at low temperatures.

Light

In a comparison of growth of loblolly pine and sweetgum seedlings in a controlled environment, Tolley and Strain (1984a) found that sweetgum far outgrew loblolly pine at a CO_2 concentration of 675 ppm at a low irradiance. Sweetgum grew 86% taller and produced 31% more leaf area and 28% more biomass at an irradiance of 250 $\mu mol\ m^{-2}\ s^{-1}$ and 675 ppm of CO_2 than at 1000 $\mu mol\ m^{-2}\ s^{-1}$ and 350 ppm of CO_2. In contrast, loblolly pine seedlings grew only 11% taller and produced less biomass and leaf surface with low irradiance and a high concentration of CO_2 than with high irradiance and 350 ppm of CO_2.

It appears that the compensatory effects of CO_2 for low irradiance may have differential effects on the success of seedlings of some species competing on the forest floor. This deserves careful study because in the long term it might cause changes in species composition of forests and other ecosystems.

Mineral Deficiencies

No data are available for woody plants, but the data from experiments with soybeans and wheat indicate that doubling the CO_2 concentration compensates at least in part for deficiencies in mineral nutrients (Sionit et al., 1981; Sionit, 1983). Patterson and Flint (1982) found that doubling the CO_2 concentration increased the growth of soybeans and several weeds, even when growth was limited by mineral deficiencies. It seems likely that the ratio of carbon to nitrogen will be higher in plants grown with a high atmospheric concentration of CO_2, although the data available are equivocal (Sionit et al., 1981; Pearcy and Bjorkman, 1983). However, it can be expected that if dry-matter production is increased the use of nitrogen and mineral nutrients also will increase. Thus, it seems safe to predict that more nitrogen and in some areas more phosphorus will be required for woody plants to obtain maximum benefits from the CO_2 increase.

Because of the complexity of these interactions, the differences among plants of various species, and the variations in weather, it is very difficult to predict the effects of CO_2 buildup on a particular crop or forest stand in a particular year (Wong, 1979; Osmond et al., 1980). As mentioned earlier, Waggoner (1984) expects relatively small changes in crop yields because the increases in dry-matter production predicted from CO_2 effects on physiological processes are likely to be counterbalanced by unfavorable climatic changes. It seems probable that for many years the short-term perturbations from normal rainfall and temperature are likely to have more effect on tree growth than the predicted climatic changes caused by the increasing concentration of CO_2 in the atmosphere.

MISCELLANEOUS EFFECTS OF INCREASING CARBON DIOXIDE CONCENTRATION

A few other effects of high CO_2 must be considered.

Flowering and Seed Production

The effects of an increasing concentration of atmospheric CO_2 on flowering and seed production are important for the natural regeneration of forests as well as for selection and breeding programs. Unfortunately, nothing is known about the effects of CO_2 concentration on the reproductive processes of trees, but if differences develop among species this might affect the species composition of forests in the long term. Zangerl and Bazzaz (1984) stated that the relative proportions of biomass partitioned into seed were affected differently in several herbaceous annuals. However, observations are needed on trees at every stage of reproduction, from flower initiation through flowering, pollination, and fertilization to the quantity and quality of seed produced. It should be possible to study the effects of CO_2 concentration on reproduction of some species by grafting branches from reproductive-age trees on seedlings growing in controlled environments.

Flowering and fruiting generally deplete the carbohydrate reserves of trees, and there is some evidence that treatments which increase the carbon-nitrogen ratio in fruit trees increase flowering (Kramer and Kozlowski, 1979). Because high CO_2 levels generally result in high ratios of carbon to nitrogen in leaves, atmospheric buildup might case earlier flowering and increased seed production. The use of enhanced CO_2 concentration for this purpose in tree-breeding experiments needs to be investigated.

Seedling Establishment and Stand Composition

Although an experimental high concentration of CO_2 has increased the rate of emergence and early growth of mountain laurel (*Kalmia*) seed (Heichel and Jaynes, 1974), it is doubtful that doubling the atmospheric concentration will have any significant effect on seed germination because the concentration already is much higher in soil than in air. However, numerous greenhouse and growth-chamber experiments indicate that after germination seedling growth is increased by the increase in leaf area and in rate of photosynthesis per unit of leaf area. This suggests that an increased concentration of CO_2 might increase growth of container-grown seedlings in greenhouses or other enclosures.

A buildup of atmospheric CO_2 is projected to change the composition of annual plant communities, especially where C_3 and C_4 plants compete (Bazzaz and Carlson, 1984). In general the competitive capacity of C_3 relative to C_4 plants should be increased by increasing concentrations of CO_2 (Strain and Bazzaz, 1983). However, all forest trees have the C_3 carbon pathway, and the differences in competitive capacity among species are likely to be smaller among trees in forests than in communities of annual plants. Thus long-term experiments spanning a decade may be necessary to demonstrate whether interspecific differences are important to foresters.

As mentioned earlier, high concentrations of CO_2 increase drought and shade tolerance, and this should increase growth of seedlings on the forest floor. If important differences in reaction to CO_2 occur among competing species, this may produce long-term changes in the composition of naturally regenerated forest stands.

Wood Quality

The effects of rising ambient CO_2 concentration on wood quality are unknown. In the only published experiment, exposure of pine and sweetgum seedlings to high concentrations of CO_2 caused no change in wood density of loblolly pine and a small increase in density of sweetgum (Rogers et al., 1983). In another unpublished experiment, there were small increases in density of wood from seedlings of both species. Data are needed from older trees.

Tolerance of Air Pollution

Increasing concentration of CO_2 is likely to decrease injury from air pollution because partial closure of stomata should reduce the entrance of pollutants into leaves. Green and Wright (1977) also reported that elevated CO_2 caused a significant increase in photosynthesis of ponderosa pine damaged severely by air pollution. Carlson and Bazzaz (1982) reported that a high concentration of CO_2 decreased injury to plants of several C_3, herbaceous species from fumigation with sulfur dioxide (SO_2).

Root Reactions

The increase in atmospheric CO_2 will have no direct effect in soil, where the concentration normally is far above 1,000 ppm. However, if the supply of carbohydrates to roots is increased this might increase mycorrhizal development and nitrogen fixation by roots of leguminous trees.

Diseases and Insect Pests

Changes in carbohydrate-nitrogen ratios of plants and changes in rainfall and temperature patterns are almost certain to change the kinds and severity of diseases and pests. For example, a drier climate and more water stress are likely to increase attacks by boring insects. In a preliminary study in the Duke University Phytotron (Lincoln et al., 1984), larvae of the soybean looper (*Pseudoplusia includens*) ate more tissue per larva on leaves of soybeans grown at 650 ppm of CO_2 than on those of plants grown in a 350-ppm atmosphere. The difference in rate of feeding was correlated with the difference in carbon-nitrogen ratio and was greater in leaves with a high carbon-nitrogen ratio. As elevating CO_2 usually tends to increase the ratio of carbon to nitrogen, the greater leaf area at higher concentrations of CO_2 might be offset by higher herbivory. This research area deserves more investigation.

ACKNOWLEDGMENTS

The research on woody plants by N. Sionit, L. C. Tolley, and B. R. Strain cited in this paper was supported by National Science Foundation grants no. DEB78-23640 and BSR82-15533. That of Rogers and his associates (1983) was supported by Department of Energy interagency grants through the U.S. Department of Agriculture.

REFERENCES

Bazzaz, F. A., and R. W. Carlson. 1984. The response of plants to elevated CO_2. I. Competition among an assembly of annuals at different levels of soil moisture. *Oecologia* (in press).

Beadle, C. L., P. G. Jarvis, and R. E. Neilson. 1979. Leaf conductance as related to xylem water potential and carbon dioxide concentration in Sitka spruce. *Physiol. Plant.* 45, 158-166.

Boyer, J. S. 1970. Differing sensitivity of photosynthesis to low leaf water potentials in corn and soybean. *Plant Physiol.* 46, 236-240.

Carlson, R. W. and F. A. Bazzaz. 1982. Photosynthetic and growth response to fumigation with SO_2 at elevated CO_2 for C_3 and C_4 plants. *Oecologia* 54, 50-54.

Cave, G. H., L. Tolley, and B. R. Strain. 1981. Effects of carbon dioxide enrichment on chlorophyll content, starch content and starch grain structure in *Trifolium subterraneum* leaves. *Physiol. Plant.* 51, 171-174.

Clough, J. M., M. M. Peet, and P. J. Kramer. 1981. Effects of high atmospheric CO_2 and sink size on rates of photosynthesis of a soybean cultivar. *Plant Physiol.* 67, 1007-1010.

Cooke, J. R. and R. H. Rand. 1980. Diffusion resistance models, p. 94-121. *In* J. D. Hesketh and J. W. Jones (ed.), *Predicting Pho-*

tosynthesis for Ecosystem Models, vol. 1. CRC Press Inc., Boca Raton, Fl.

Downton, W. J. S., O. Bjorkman, and C. S. Pike. 1980. Consequences of increased atmospheric concentrations of carbon dioxide for growth and photosynthesis of higher plants. *In* G. I. Pearman (ed.), *CO_2 and Climate, Australian Research.* CSIRO, Australian Academy of Science, Canberra, Australia.

Gifford, R. M., J. H. Thorne, W. D. Hitz, and R. T. Giaquinta. 1984. Crop productivity and photoassimilate partitioning. *Science* 225, 801-808.

Green, K., and R. D. Wright. 1977. Field responses of photosynthesis to CO_2 enhancement in ponderosa pine. *Ecology* 58, 687-691.

Heichel, G. H., and R. A. Jaynes. 1974. Stimulating emergence and growth of *Kalmia* genotypes with carbon dioxide. *Hortscience* 9, 60-62.

Hofstra, G., and J. D. Hesketh. 1975. The effects of temperature and CO_2 enrichment on photosynthesis in soybean, p. 71-80. *In* R. Marcelle (ed.), *Environmental and Biological Control of Photosynthesis.* J. Junk, The Hague.

Kaufmann, M. R. 1984. Em-CWU, a canopy model for determining transpiration of subalpine forests. I. Model development. *Can. J. For. Res.* 14, 18-226.

Kimball, B. A. 1983. Carbon dioxide and agricultural yield: an assemblage and analysis of 430 prior observations. *Agron. J.* 75, 779-788.

Kramer, P. J. 1981. Carbon dioxide concentration, photosynthesis, and dry matter production. *BioScience* 31, 29-35.

Kramer, P. J. 1983. *Water Relations of Plants.* Academic Press, Inc., New York.

Kramer, P. J., and T. T. Kozlowski. 1979. *Physiology of Woody Plants.* Academic Press, Inc., New York.

Kriedemann, P. E., R. J. Sward, and W. J. S. Downton. 1976. Vine response to carbon dioxide enrichment during heat therapy. *Aust. J. Plant Physiol.* 3, 605-609.

LaMarche, V. C., Jr., D. A. Graybill, H. C. Fritts, and M. H. Rose. 1984. Increasing atmospheric carbon dioxide: tree ring evidence for growth enhancement in natural vegetation. *Science* 225, 1019-1021.

Lincoln, D. E., N. Sionit, and B. R. Strain. 1984. Growth and feeding responses of *Pseudoplusia includens* (Lepidoptera: Noctoidae) to host plants growth in controlled carbon dioxide atmospheres. *J. Environ. Entomol.* 13, 1527-1530.

Miller, E. C. 1938. *Tree Physiology, 2nd ed.*, p. 458-459. McGraw-

Hill, New York.

Nobel, P. S. 1980. Leaf anatomy and water use efficiency, p. 43-55. *In* N. C. Turner and P. J. Kramer (ed.), *Adaptation of Plants to Water and High Temperature Stress*. Wiley, New York.

Osmond, C. B., O. Bjorkman, and D. J. Anderson. 1980. Physiological processes in plant ecology. Ecological Studies 36. Springer-Verlag, New York.

Patterson, D. T. and E. P. Flint. 1982. Interacting effects of CO_2 and nutrient concentration on soybean (*Glycine max*), sicklepod (*Cassia obtusifolia*), and showy crotolaria (*Crotolaria spectabilis*). *Weed Sci.* 30, 389-394.

Pearcy, R. W., and O. Bjorkman. 1983. Physiological effects, p. 83, 84. *In* E.R. Lemon, (ed.), *CO_2 and Plants--the Response of Plants to Rising Levels of Atmospheric Carbon Dioxide*. Westview Press, Boulder, Colo.

Regehr, D. L., F. A. Bazzaz, and W. R. Boggess. 1975. Photosynthesis, transpiration and leaf conductance of *Populus deltoides* in relation to flooding and drought. *Photosynthetica* 9, 52-59.

Rogers, H. H., G. E. Bingham, J. D. Cure, J. M. Smith, and K. A. Surano. 1983. Responses of selected plant species to elevated carbon dioxide in the field. *J. Environ. Qual.* 12, 569-574.

Sharp, R. E., and W. J. Davies. 1979. Solute regulation and growth by roots and shoots of water-stressed maize plants. *Planta* 147, 43-49.

Sinclair, T. R., C. B. Tanner, and J. M. Bennett. 1984. Water use efficiency in crop production. *BioScience* 34, 36-40.

Sionit, N. 1983. Response of soybean to two levels of mineral nutrition in CO_2-enriched atmosphere. *Crop Sci.* 23, 329-333.

Sionit, N., H. Hellmers, and B. R. Strain. 1982. Interaction of atmospheric CO_2 enrichment and irradiance on plant growth. *Agron. J.* 74, 721-725.

Sionit, N., and P. J. Kramer. 1984. Woody plants reactions to carbon dioxide enrichment. *In* H. Z. Enoch and B. A. Kimball (ed.), *Carbon Dioxide Enrichment of Greenhouse Crops*. CRC Press, Boca Raton, Fl. (in press).

Sionit, N., D. A. Mortensen, B. R. Strain, and H. Hellmers. 1981a. Growth response of wheat to CO_2 enrichment and different levels of mineral nutrition. *Agron. J.* 73, 1023-1027.

Sionit, N., B. R. Strain, and H. A. Beckford. 1981b. Environmental controls on the growth and yield of okra. I. Effects of temperature and of CO_2 enrichment at cool temperature. *Crop Sci.* 21, 885-888.

Sionit, N., B. R. Strain, H. Hellmers, and P. J. Kramer. 1981c. Effects of atmospheric CO_2 concentration and water stress on water relations of wheat. *Bot. Gaz.* 142, 191-196.

Sionit, N., B. R. Strain, H. Hellmers, G. H. Riechers, and C. H. Jaeger. 1985. Long term atmospheric CO_2 enrichment affects the growth and development of *Liquidambar styraciflua* L. and *Pinus taeda* L. seedlings. *Can. J. For. Res.* (in press).

Strain, B. R., and F. A. Bazzaz. 1983. Terrestrial plant communities, p. 179, 193. *In* E. R. Lemon (ed.) *CO_2 and Plants--The Response of Plants to Rising Levels of Atmospheric Carbon Dioxide.* Westview Press, Boulder, Colo.

Tolley, L. C. 1982. The effects of atmospheric carbon dioxide enrichment, irradiance and water stress on seedling growth and physiology of *Liquidambar styraciflua* and *Pinus taeda.* Ph.D. dissertation. Duke University, Durham, N.C.

Tolley, L. C. and B. R. Strain. 1984a. Effects of CO_2 enrichment on growth of *Liquidambar styraciflua* and *Pinus taeda* seedlings under different irradiance levels. *Can. J. For. Res.* 14, 343-350.

Tolley, L. C., and B. R. Strain. 1984b. Effects of atmospheric CO_2 enrichment and water stress on growth of *Liquidambar styraciflua* L. and *Pinus taeda* L. seedlings. *Can. J. Bot.* (in press).

Waggoner, P. E. 1984. Agriculture and carbon dioxide. *Am. Sci.* 72, 179-184.

Wong, S. C. 1979. Elevated atmospheric partial pressure of CO_2 and plant growth. I. Interactions of nitrogen nutrition and photosynthetic capacity in C_3 and C_4 plants. *Oecologia* 44, 68-74.

Wulff, R. D., and B. R. Strain. 1982. Effects of CO_2 enrichment on growth and photosynthesis in *Desmodium paniculatum. Can. J. Bot.* 60, 1084-1091.

Zangerl, A. R. and F. A. Bazzaz. 1984. The response of plants to elevated CO_2. II. Competitive interactions among annual plants under varying light and nutrients. *Oecologia* (in press).

HOW FOREST PRODUCTS COMPANIES CAN RESPOND TO RISING CARBON DIOXIDE AND CLIMATE CHANGE

by Robin Sandenburgh, Carol Taylor, and John S. Hoffman[1]

ABSTRACT

Forest products companies will have to decide how to respond to possible climate change. One option is to do nothing; another is to make consideration of climate change a part of every operational decision. Companies may want to consider action which would enhance profits if climate changes. Decisions fall into two categories: (1) resource acquisition and management; and (2) corporate research and development decisions. However, industry lacks the information necessary to make these decisions.

Industry could begin by assessing the value of timely responses. Lacking adequate information about carbon dioxide (CO_2) buildup, firms will want to know more about the value of accelerating their search for response strategies. There are obstacles: difficult economic conditions, lack of agreement on facts, institutional inattention, a feeling that it is someone else's problem, and complacency about the present state of knowledge.

Nonetheless, timely responses could result in significant benefits: higher profits and stock prices, a long-term competitive advantage, and reduced risk levels.

There are a number of strategic and operational decisions which might be made. Strategically, forest products companies could improve their position through acquisition of land, development of infrastructure, and new approaches to marketing. Operationally, they might assess planting and reforestation and timing of timber harvests.

Decisions in the forest products industry must be made in the face of considerable uncertainties about future resource and market conditions. Strategic planning seeks to identify profitable opportunities and to avoid potentially damaging risks despite those uncer-

[1]This material is adapted from the report, *Strategic Options to Enhancing Forest Industry Productivity in the Face of Rising Atmospheric Carbon Dioxide*, prepared by ICF Incorporated for the U.S. Environmental Protection Agency.

tainties. The rising concentrations of atmospheric "greenhouse" gases and carbon dioxide (CO_2) present new challenges in future strategies for the forest products industry.

Forest products companies will have to decide how, if at all, to respond to possible changes in climate. Choices range from doing nothing now and hoping that any changes can be dealt with without significant losses, to making climate change and rising carbon dioxide part of every strategic and operational decision. Deciding what actions to undertake between these two extremes requires analyzing vulnerabilities and opportunities to determine whether profits could be sufficiently enhanced to warrant the effort of adding these concerns to the already complex decision making process.

The ability to "survive" the change in climate is not the real issue. The net effects of climatic and biochemical change on productivity may be beneficial, even if current decisions do not anticipate changes. Most forest products firms, however, seek more than survival; they want to increase profits and productivity to the highest level compatible with prudent risk taking. Firms may wish to consider the changes that could occur and to analyze which options would enhance profits if predicted shifts in climate and plant biology become manifest.

Two very different types of decisions need to be considered:

- *Resource acquisition and management decisions.* Firms can alter the tree species planted, management practices employed, land acquired or sold, and supporting infrastructure developed.
- *Corporate research and development decisions.* Firms can assess the value of accelerating research to identify CO_2 effects, develop options, and understand changes to make future investments profitable.

Even if corporate decision makers accept climate change and increasing atmospheric CO_2 as an important issue, current information has not yet been sufficiently analyzed for them to develop adequate responses for these kinds of decisions. Existing information is insufficient to determine the extent to which resources need to be shifted to respond to the possibility of climate change. For example, no estimate has been made of the value of engaging in more scientific research to reduce uncertainties more quickly; the options and costs of doing so have not even been studied. Nor has an estimate been made of the extent to which productivity and profits could be enhanced by changes in current decisions. Until development of such information begins in earnest, corporate managers will have little basis for choosing the most appropriate responses. In light of this, forest managers may want to develop their response strategies over time, using results from short-term projects to select areas for further analysis and to target investments that seem most likely to pay off.

ASSESSING THE VALUE OF TIMELY RESPONSES

Systematically analyzing decisions requires the expenditure of resources. Firms, therefore, may wish to review the decisions that may be affected in phases, identifying and considering the highest priorities first.

The earlier the most productive possibilities are identified, the more profitable they are likely to be. For example, new information might indicate that certain species are more productive than others under conditions of higher atmospheric CO_2. If planting strategies were altered on the basis of this new information, the value of that decision would be the sale price of the added timber, appropriately discounted to account for the time value of money. To reduce costs, firms may wish to assess just how much a knowledge of potential climate change and CO_2 effects would be worth before undertaking experiments, just as they would probably wish to undertake experiments before changing planting decisions.

The possibility of realizing benefits from a particular planting, protection, harvesting, acquisition, or research decision cannot, by itself, justify a decision that requires capital expenditure. Lacking adequate information about CO_2 buildup, its effects, and possible opportunities for enhancing productivity, firms in general will wish to know more about the value of accelerating their search for appropriate strategies.

Decision analysis can assist in calculating the value of obtaining critical information earlier rather than later (Raiffa, 1970). Decision analysis has been used in a variety of fields, including forestry. A decision analyst first examines how productive different decisions would be, under a range of plausible scenarios about future conditions, and second, evaluates the increase in profits that would occur if the firm knew *in advance* which conditions would develop. By summing the added profits that would accrue over the lifetime of the decision, an estimate of the value of earlier information can be obtained.

Many well-developed quantitative methods now exist for applying decision analysis. Its tools can estimate the long-term value of a particular forest management decision over time, under different assumptions about the information available to the firm at the time the decision must be made. Estimates of the long-term value of a particular holding could be based on computerized models of forest productivity, which calculate future forest output under various environmental, financial, and silvicultural conditions (Duerr et al., 1979; Trimble and Shriner, 1981). Such models calculate the growth achieved in each year and total these estimates over the life of the stand. To determine a stand's value today, the estimate of harvestable yield is multiplied by the expected price for the output, discounted appropriately.

OBSTACLES TO DEVELOPING A STRATEGIC RESPONSE

If developing a response strategy early could significantly increase yield and profits, forest products firms would automatically add climate change and the CO_2 issue to their strategic agenda. However, certain institutional and political impediments exist that must be overcome if CO_2 buildup is to be considered.

Current Economic Conditions

Recent economic conditions have forced many industries to take a cautious stance on investments and decisions. A danger exists, however, in being too conservative. Caution may be inappropriate because efforts need not require large commitments of time or resources. In the case of CO2-related projects, simply raising the issue and incorporating it into the firm's ongoing planning process may be a good start.

Sorting Out the Facts

Scientific exchange, by its very nature, tends to focus on disagreements rather than consensus. Frequently, differences between scientists obscure important facts. To many, the mere publication of dissent indicates uncertainty. To nonspecialists, scientist and nonscientist alike, there sometimes appear to be major disagreements among climatologists when, in fact, none exists. Understanding the scientific debate among experts is difficult. Distinguishing those who have the breadth of experience or expertise to make valid conclusions from those who do not is never easy. And, as often happens, when scientists of specialized fields are asked to comment on subjects outside their areas of expertise, casual comments lead to misinterpretations.

Institutional Inattention

No "Division of Climate Change and CO_2 Effects" exists in forest products companies. The changes that could be induced by rising CO_2 cut across many activities and could potentially influence dozens of functions within a firm. Yet individual managers may deem these complex changes beyond their purview. Furthermore, departmental entrenchment--committed participants and obligated budgets--may bar the climate change and CO_2 issue. Particularly in a time of general economic stress, the likelihood of CO_2-related research "falling through the cracks" is strong. To negate these pressures, CO_2-related concerns must be elevated to a position warranted by their potential economic importance. Delays in responding may prove costly, especially for those research areas in which experiments cannot later be compressed into shorter time frames.

"It's Someone Else's Problem"

Because rising atmospheric CO_2 levels and their effects could influence many other industries in addition to forest products, a tendency to wait for others to do something about the situation is tempting. Logically everyone could succumb to this temptation. If the importance of learning more about CO_2 effects is not recognized in individual industries, the overall value of learning more earlier can never be assessed. Valuable time and resources will be lost as everyone waits for someone else to start the analysis.

Complacency About Knowledge

If existing geographic variation is mistaken as equivalent to the variability that CO_2 buildup could induce in particular locations, corporations could be lulled into thinking that existing information is adequate to deal with historically unprecedented conditions. Forest growth depends on climatic and ecological conditions over the lifetime of a stand. Climate and ecological conditions have always varied from place to place and over time; forest managers therefore have to make decisions based on some degree of variability. However, if the projections of some scientists for greenhouse gas-induced climate change are correct, a forest stand, over its life, could face total environmental variability beyond any existing today. Therefore, a firm's competence to deal with regional differences and changes within a region should not be mistaken as knowledge of how to respond to an entirely different sort of change and variation, for which historical precedents and existing responses may be inadequate.

ADVANTAGES OF APPROPRIATE RESPONSES

In deciding on a firm's strategic responses, corporations will want to evaluate whether possible benefits will outweigh the costs and difficulties of considering the issue.

Higher Profits and Stock Prices

Decisions that anticipate future conditions accurately are likely to raise long-term profitability. As conditions change, the superiority of certain decisions will become apparent, and resulting higher profits will be anticipated by market analysts. Anticipated profits generally have immediate effects on stock prices. Thus, even though profits resulting from better decisions may not be realized for a long time, their value would be captured as soon as the market recognizes the wisdom of these decisions. Firms that do not alter decisions and that fail to adapt to future conditions could suffer, losing both product and profitability.

Permanent Competitive Advantage

Creating an information base and a technological capability to respond to future conditions--in this case increased atmospheric CO_2--induces superior development of useful information and appropriate technologies. Adapting to possible CO_2-induced changes may proceed along a "learning curve," in which adaptation becomes easier and more effective as time goes on and capabilities grow. Firms that gain a first advantage may maintain or enhance their lead over time. Advantages of this sort may be permanent, such as the development of a superior genotype or acquisition of exceptionally productive forestlands.

Reduced Risk Levels

Firms generally seek to increase profits only within the bounds of acceptable risk; failure to comprehend changes in the environment could subject a firm to greater financial risk than is strategically desirable. If rising CO_2 changes the world's climate, the mix of risks facing firms would change considerably, too. Preparing for possible CO_2-induced changes could improve a firm's ability to manage risk.

Escalating Importance

In making the assessment of how fast to proceed, forest products firms need to consider one important likelihood: If the greenhouse hypothesis is correct, each decade will bring changes in the climate and biosystem greater than those occurring in the decade before, thereby constantly increasing the relative importance of understanding and adapting to these changes. Obviously, decisions based on an assumption that future growing conditions will be like those of the past become less and less appropriate for achieving the highest possible yield. Thus, the earlier better information is available to improve planting, harvesting, and investment decisions, the greater the possibility of maintaining or enhancing productivity.

A FRAMEWORK FOR ANALYSIS

In light of the range of effects that increased atmospheric CO_2 buildup could have on forests' growing conditions, forest products firms may wish to consider how, and at what rate, their response actions can be defined, analyzed, and implemented. One approach to this problem is pinpointing changes in specific decisions and other corporate activities necessary to take into account CO_2-induced changes. Firms will want to respond to the emerging CO_2 issues differently, depending on their particular characteristics and needs. The following is an analytical framework and process for assessing developments as they emerge.

Strategic Decisions

A forest products firm's profits depend on acquiring and developing the resources best suited to future growth and market conditions. A number of decisions could be influenced by CO_2-induced changes. Some decisions, such as those dealing with land acquisitions, must be made long before changes occur. Others, such as fire prevention strategies, can wait until change requires action.

Purchase of Productive Forestlands

The decision to purchase or expand production to new land is an important one because the productive potential of a firm's land limits its possible profitability. Identifying areas in which CO_2--induced climate and environmental change might improve or worsen existing productive capacity is of major importance to forest products firms. Lands on which future conditions may become less favorable can be avoided, and lands on which future conditions may be productive can be purchased. Specific actions that could be taken now include:

- establishing purchase options in areas that may increase in comparative productivity over the next several years;
- diversifying holdings geographically to reduce exposure to unexpected negative changes in particular regions; and
- selling lands in high-risk areas.

Infrastructure Development

The second major investment decision encountered by corporate planners is the location of important infrastructure elements, such as roadways or other forms of transportation and new timber-processing mills. An important criterion is that the mill be located in the proximity of a predictable source of timber. Clearly, the expected effects of rising CO_2 levels on timber sources could be important. Options to address these concerns include:

- considering likely future CO_2 effects on specific regional growing conditions and timber supplies; and
- analyzing possible changes in the flow conditions of rivers and streams when these changes would influence transportation options.

Insurance Needs

Insurance coverage needed to protect against catastrophic losses may change with increased CO_2. The level of atmospheric CO_2 may influence the types of risks involved in forest management and their distribution.

Basic Marketing Decisions

Decisions on which markets to enter, expand, avoid, or exit depend on the expected profitability of each market. While it is impossible at this time to determine whether CO_2-induced alterations will change national or international productivity sufficiently to affect the availability of different products, the possibility exists that such a change could occur in the future. Firms may wish to develop and consider information useful in analyzing changing markets and making market decisions.

Operational Decisions

Silvicultural management entails making decisions and choosing investments to best manage a stand or holding. CO_2-induced changes are likely to be important in three distinct stages in the management of a stand or holding: (1) planting/reforestation; (2) pest protection; and (3) harvesting. At each of these stages, decisions can be changed now and in the future to contend with both the potential influences of CO_2 and the added elements of uncertainty.

Planting / Reforestation

Once a stand has been established, the only methods for preventing productivity losses or enhancing gains, if conditions change from those initially predicted, might be: (1) to harvest early on, or (2) to take steps that are labor- and cost-intensive. Anticipating changes in climate and environment, or even a range of possible changes, may allow companies to plant species that benefit the most from positive effects, and/or are less vulnerable to negative effects, or to plant a range of species with a variety of characteristics, such as shorter maturation cycles or lower water requirements, thereby increasing product diversity and lowering total risk.

Pest Protection

Pest protection efforts encompass many activities developed to make forest stands more productive, including chemical, biological, and silvicultural methods. Climate shifts and the resulting changes in the pest environment could necessitate altering the overall protection strategy for a given region, since CO_2-induced changes may alter both the type and magnitude of the pest threats. Types of actions that could be taken include developing monitoring capabilities to identify new pests before they spread; accelerating harvest in areas of new pest invasion or of rapid growth in the population of an existing pest; investing in greater "reserve" capacity for fighting pests; and increasing associations with chemical firms, computer companies, and academic scientists to deploy the best integrated pest management system under various climate scenarios.

Fire Control and Prevention

Shifts in temperature and rainfall could increase the probability and severity of forest fires in some areas and necessitate a greater investment in fire control and prevention techniques. Measures worth considering include cutting earlier and more frequently to create more firebreaks, acquiring additional equipment reserves, and purchasing more insurance coverage.

Timing of Timber Harvests

Decisions about the timing and extent of the yearly harvest are complex, because many factors must be considered. In general, the annual growth rate increases over time, peaks, and then declines until the death of the tree. To maximize the volume of wood grown per acre, trees are harvested when the tree has reached the peak of its mean annual increment (MAI). Since CO_2 levels influence growth rates in several ways, the age-growth curves for each tree species could shift as the level of CO_2 in the atmosphere increases. The result of such shifts will be that traditionally optimal harvest times could change. Thus, increasing ambient CO_2 could affect both the present and future financial components of the decision, because of the pervasive influence of atmospheric change on the shape of future age-growth curves.

Another criterion for harvest is financial maturity, when the margin of income over costs for a particular tree species is maximized. Because of the financial consideration, the optimal harvest point often precedes maximum MAI. CO_2 levels would affect this decision rule as well, because the cost of inputs may be affected by CO_2-related environmental changes, and the amount of annual growth could change.

CONCLUSION

The principal decision that forest products companies must make in the face of possible changes in climate due to the predicted rise of atmospheric CO_2 is whether to do something about it or not. If a firm makes the decision to respond to the challenge in an attempt to maximize benefits and minimize potential losses, it must carefully analyze its options both in resource management and corporate strategy. A number of obstacles impede such analysis, including insufficient information, financial considerations, and corporate inertia. However, appropriate responses in both the short and the long run could result in benefits to the forest products industry in the form of higher profitability, reduced risks, and a continuing and possibly increasing competitive advantage in the years to come.

REFERENCES

Duerr, W. A., et al. 1979. *Forest Resource Management: Decision Making Principles and Cases.* W.B. Saunders & Co., Philadelphia.

Raiffa, H. 1970. *Decision Analysis.* Addison-Wesley, Boston.

Trimble, J. L., and C. R. Shriner 1981. *Inventory of United States Forest Growth Models.* Oak Ridge National Laboratory Report ORNL/Sub-80/13819/1. U.S. Department of Energy, Washington, D.C.

A CONCEPTUAL FRAMEWORK FOR ASSESSING IMPACTS OF CARBON DIOXIDE CHANGE ON FOREST INDUSTRIES

by Dietmar W. Rose, Alan R. Ek, and Keith L. Belli

ABSTRACT

An analytical framework was developed for assessing possible impacts on forest industries of global warming caused by rising atmospheric carbon dioxide (CO_2) levels. Impacts on the aspen-based forest industry of the Great Lakes states and the loblolly pine-based forest industry of the southeastern United States were analyzed with available, limited information.

In the unchanged-CO_2 scenario, long-term supply shortages are predicted for the Great Lakes aspen-based and the southeastern United States loblolly pine-based industries. These shortages can be reduced through technological changes and substitution with underutilized hardwood species. Because an increase of the aspen and loblolly pine range is expected under the doubled-CO_2 scenario, industrial decisions to deal with pending supply shortages will be influenced more by less speculative factors than future climate. Changes in species ranges will, however, have important implications for state forestry programs.

Rising amounts of atmospheric carbon dioxide (CO_2) could lead to substantial increases in average global temperatures and to dramatic climatic changes (Rind and Lebedeff, 1984). Specific regional changes in climatic conditions such as rainfall amount and distribution, length and severity of droughts, number of frost-free days, and temperatures could clearly affect forest growth and the forest products industry. Unfortunately, it is far too early and too much information is still lacking to accurately quantify possible biological, economic, and social impacts of possible global warming trends on the forest products industry. It is not too early, however, to organize available information to identify major gaps in knowledge and to develop a conceptual framework for analysis.

FORMULATING QUESTIONS

According to Hoffman (1984), no single environmental or technical change is likely to have a more pervasive influence on forestry than rising amounts of atmospheric CO_2. To be appreciated, however, this statement requires specification of a frame of reference. If the projected changes in CO_2 and associated climatic changes occurred overnight, the consequences could be catastrophic. More likely they will occur over decades. In any case, if climatic changes resulting from increasing CO_2 could be predicted with reasonable accuracy, then appropriate steps to minimize the negative and take full advantage of the positive aspects could be taken.

The most likely scenario of future atmospheric CO_2 buildup is that changes will occur continually and that the development of appropriate strategies needs to include considerable flexibility to respond to dynamic changes. The forest industries have the ability to respond to these changes; they are used to making decisions in response to changing markets, technology, competition, and silvicultural know-how. An example is the massive move of forest industries from the West Coast to the South in response to production cost differentials, reliability of timber supplies, and proximity of consumer markets. Another example is the rapid emergence of the waferboard industry in the Great Lakes states in response to high prices of peeler logs for plywood on the West Coast, large and available supplies of aspen in the Great Lakes states, and a new technology for producing a substitute product. The spatial and time dimensions of forestry complicate impact analysis considerably. Changes in temperature, water availability, and other climatic variables need to be known with respect to timing and geographical distribution. Only then can questions be formulated about changes in the ecology of forests and related changes in growth, mortality, and susceptibility to pests and fire. The following list indicates the kinds of facts needed to begin an impact analysis:

- climate and "weather" by locality over time;
- silvicultural characteristics of species;
- response of individual species to increased CO_2 and associated climatic changes;
- technology of species utilization, particularly future technology;
- distribution of covertypes by age, volume, owner class, and location;
- location, capacity, and developmental stage of forest indus-

tries;

- supply schedules over time; and
- markets and demands over time.

FRAMEWORK FOR IMPACT ANALYSIS

Because the impacts of CO_2 changes will be to some degree a function of factors directly under the control of a forest industry company, it is useful to describe these factors or decision variables. Decision variables are determined by a number of considerations, and the choices among alternative decisions are often affected by factors outside the control of the company. Figure 1 illustrates a model of this strategic decision process.

Issues and Considerations

Corporate decision making has always included the issues listed in Figure 1, except for potential climatic changes. To incorporate this issue into the decision-making process, a state-of-the-art summary of the biological and technical aspects of species and varieties under various climate scenarios needs to be developed.

This summary might include maps that show potential distribution of each species under various climatic scenarios hinging on CO_2 increases. Additionally, this would require matching environmental ranges of individual species with present and predicted future climatic maps. The study of existing species distribution maps with respect to climate and overlays of current industrial centers and consumer markets can provide a quick reference to where major changes might occur and where major attention needs to be paid to developing strategies to reduce the negative impacts of climatic changes. It is similarly important to assess the effect of changes in climate on agricultural plant ranges (that is, corn, wheat, and soybeans), because these may compete with forestry for available land.

Decision Variables

The major areas for consideration by forest industries are species selection, rotation length, management inputs, land ownership, and research and development.

Species Selection

At the time of reestablishing a forest site by natural regeneration

Figure 1

Issues / considerations	Decision variables	Decision-making support models	Strategies
Climatic changes	Species selection	Growth simulation	Long-term:
Spatial distribution of forests, production facilities, markets	Rotation length	Cash-flow analysis	---switch species
Technology	Management inputs	Harvest scheduling	---switch to other location
Competition	Ownership of land	Supply analysis	--get out of business
Land use / ownership	Research and development	Demand analysis	---expand ownership
Demand / supply			Short-term:
			---buy additional wood at higher prices
			---reduce production capacity

Components of the strategic decision process for the forest industry.

or by planting, a firm commits capital for as long as one rotation. The selection of species is influenced by the predicted demand for products manufactured from these species. Predicted demands will influence decisions on expanding processing capacity. Under conditions of tight supplies, companies will look at ways to increase supplies by buying additional land and by developing technologies that substitute other species in the production process. Climatic changes and their impact on the performance of selected species become an additional consideration in reforestation decisions. A species planted today might turn out to be a good choice for only half the rotation, after which an alternative species might become preferable for that environment if the climate changes. Because uncertainty surrounds future demand, technology, and climate, flexibility and risk-reduction need to be incorporated in reforestation decisions (for example, mixed-species stands or at least a mixture of species in the supply).

Rotation Length

A number of considerations enter the decision on rotation length. Among them are desired end product, economic returns, and biological and environmental considerations. Recent trends toward shortening of rotations result from new production techniques that better utilize trees and that do not require large-dimension timber. These new techniques also reduce the cost of production. This trend will likely continue, especially in processing innovations and improvements in tree breeding. Earlier-than-planned harvest of a plantation is another way to make a midcourse correction of a reforestation decision that, in hindsight, seems inappropriate.

Management Inputs

These inputs encompass all decisions on the type and intensity of managing established stands--for example, decisions on competition control, timing, and intensity of thinnings; on fertilization; and on the type of harvesting system. A trend to full-tree harvesting, for example, would substantially increase recoverable harvest volumes over traditional harvesting systems.

Ownership of Land

Through purchase of timberland, companies attempt to control their timber supply. Companies will sell land to move to other areas, to get out of the business, or to change emphasis to different species and products. Such decisions can be a dramatic adjustment to changing conditions of markets and climates. O'Laughlin and Ellefson (1982), in a survey of the wood-based industry, identified timber resources or land ownership as the key corporate strategy. The 50 largest wood-based firms own more than 85% of the industry's woodlands. An assured physical supply of timber for highly capital-intensive production facilities is the main reason for these holdings. Land ownership further reduces vulnerability to short-run fluctuations in stumpage prices and improves the flow of raw materials to processing facilities.

Research and Development

In the past, technological breakthroughs in the forest industry have changed, often dramatically, the structure and competitiveness of sectors of the industry between and among regions. The development of the waferboard technology transformed the aspen resource of Minnesota from what was previously considered a relatively useless weed species to a highly desirable source of raw material virtually overnight. In the area of pulping, major breakthroughs are now permitting effective use of many species. For the future, biotechnology offers the promise of developing plans able to adapt to adverse environmental conditions, to better resist diseases, and to fix nitrogen. This research will yield plants able to do well under various climatic conditions. The wood-processing companies, thanks to biotechnology research, may also develop substantial new product opportunities, such as for the silvichemical industry.

Decision-making Support Models

The strategy for a thorough impact analysis is to utilize all available data and models to analyze as many components of the complex puzzle as possible, and then to organize the findings for a rational, integrative examination. For selected climatic scenarios and assumed management alternatives, the following major analyses need to be considered: (1) growth simulation--species distribution and growth; (2) cash-flow analysis; (3) supply analysis; and (4) demand analysis.

A specific management strategy must be assumed for each scenario. In reality, no one management strategy will be best for the future because companies react to changing conditions of markets, competitors, and climate--if the latter generates visible impacts.

The physical supply of raw materials over time is the key informational item in supply, demand, and harvest-scheduling analysis. The first and most important component of an impact analysis is estimating forest growth over time under the most likely climatic scenarios. Powerful growth simulation programs are available for most major timber species, but, unfortunately, most such projection models need modifications to reflect climatic changes because climate variables rarely have been included as a model component. Growth simulations, combined with removal of stand volumes to meet demands, can point to inadequacies of local or regional timber supplies. If such shortfalls in supply can be demonstrated under normal climatic conditions, it will be even more crucial to evaluate inventory changes under changing climatic conditions.

Most likely, impacts will cover large geographical regions, and the potential of displacing ideal growing regions for agricultural and forestry crops could mean serious economic imbalances and losses. For example, a massive dieback of certain species caused by climatic changes is a possibility because ecological thresholds for some species are very sensitive (E. I. Sucoff, personal communication).

Spatial considerations are crucial for assessing impacts of CO_2 increases on forest industries. Market centers as well as production facilities are fixed, as is much of the infrastructure, especially transportation networks, railheads, and rails. A reduction in regional timber supply--regardless of cause--would affect regional competitors somewhat equally; local markets will absorb the impact through higher prices. Competitors with a higher degree of self-sufficiency will increase their competitiveness and force competitors out of the region. If the effect on supply reaches a point that the region as a whole loses its competitiveness with other regions, even the best regional competitors will close plants.

Strategies

Once the environment for decision making is described and incorporated into various decision models, strategy options can be weighed.

Strategy is the major source of organizational direction and cohesiveness. According to Nadler (1978), organizational strategy needs to consider: corporate mission or purpose; financial measures of effectiveness in achieving corporate purposes; products and costs; technological capability; sales growth and share of market; distribution strategies; acquisitions and divestitures; organizational structure; productivity improvement; environmental scanning; human resources; compensation; manufacturing facilities and policies; and pollution and environmental impact.

Strategic planning as a managerial process can be summarized as setting objectives, studying the environment, appraising strengths and weaknesses, and formulating strategies (O'Laughlin and Ellefson, 1982). Corporate strategy depends on three parameters: (1) acreage held in fee ownership; (2) timberland market values; and (3) the degree to which the company satisfies its raw material needs from its own lands.

APPLICATION

While much information needed to carry out a complete impact analysis for any part of the country is lacking, available information was used to illustrate an impact analysis of the Upper Great Lake states and the southeastern United States.

Some information on physiological responses of plants to rising atmospheric CO_2 is available. According to Leverenz and Lev (this volume), a doubling of CO_2 more than compensates for temperature increases of up to 10°C (18°F). The increase in photosynthetic efficiency resulting from higher temperatures will be somewhat reduced by increased respiration. A small increase in forest productivity is expected because of increased efficiency of light utilization under higher atmospheric CO_2 concentrations. Photosynthetic water-use efficiency is expected to at least double with a doubling of atmospheric CO_2. This factor should counteract decreases in precipitation associated with CO_2 changes. This information provides some basis for assessing reactions of species to assumed climatic changes in specific locales. However, potential insect and disease effects are not yet well understood.

Climate-change scenarios, based on projected doubling of atmospheric carbon dioxide levels, allow speculation on possible geographical range modifications of important timber species. Quantitative estimates of changes in specific climate parameters were provided by the global climate mode (GCM) used by Rind and Lebedeff (1984). For illustrative purposes, two major timber-producing regions, the Upper Great Lakes and the Southeast states, and their two most important species, aspen and loblolly pine, respectively, were chosen.

Considerable knowledge has been accumulated for the Great Lakes states' aspen resource. Growth projections of Minnesota's total timber inventory are available to the year 2030. Present and future information on spatial distribution of any inventory category is furthermore available through a sophisticated transportation model (Williams et al., 1982). Based on this model, information on hauling distances, delivered-wood cost, and relative indexes of competition can be derived readily for any existing or new market now or in the future (Rose et al., 1983 and 1984). The "unchanged CO_2" columns in Table 1 summarize major findings of a number of supply-demand studies of aspen (Smith, 1982; Williams et al., 1982; Montzka and Rose, 1982; Erkkila et al., 1982). In the past, aspen growth has far exceeded removals. The recent installation of several new waferboard plants in Minnesota has increased demand for aspen substantially and will contribute much to the rapid closing of the gap between growth and removals. Cost of available supply will increase for all major markets, over time, in line with increasing average haul distances. Allocation of available, harvestable aspen timber will increase from 59% to 93% by the year 2000. Any developing shortfalls can be covered by other vastly underutilized hardwoods, given further technological improvements. A substantial additional volume of timber is available in the form of aspen stand residues. Currently, only 39% of total hardwood biomass is removed in a pulpwood harvest. An additional 20% of volume could be recovered even with traditional logging methods.

Within the Upper Great Lakes region, projections by the GCM show increased average winter temperatures and a lengthening of the growing season by as much as a month. Changes in plant-water relationships, however, are not as simple to predict. An overall increase in precipitation is offset by higher projected evapotranspiration and runoff rates. Further complications arise from the physiological ability of plants to cope more easily with water stress, given higher partial pressures of atmospheric carbon dioxide. Assuming the GCM to be correct, the anticipated extension of the potential aspen range would be slightly westward past the Minnesota border in response to increased precipitation and enhanced drought tolerance (Figure 2).

A similar summary table was developed for loblolly pine (Table 2) to summarize the current understanding of various issues. Substantial loblolly pine acreages in the middle age-classes are the result of the massive plantation programs of the 1950s. However, inventories have shown a major decline in the small-diameter trees (2 to 4 inches, or 5 to 10 cm). Because of increasing costs, many stands have not been regenerated after harvest. This is most evident on small, nonindustrial private ownerships. Nonindustrial, private owners own about 75% of the wood and account for 85% of wood removal. Rapid inventory accumulation and fast growth still provide a comfortable margin between growth rates and removals

Figure 2

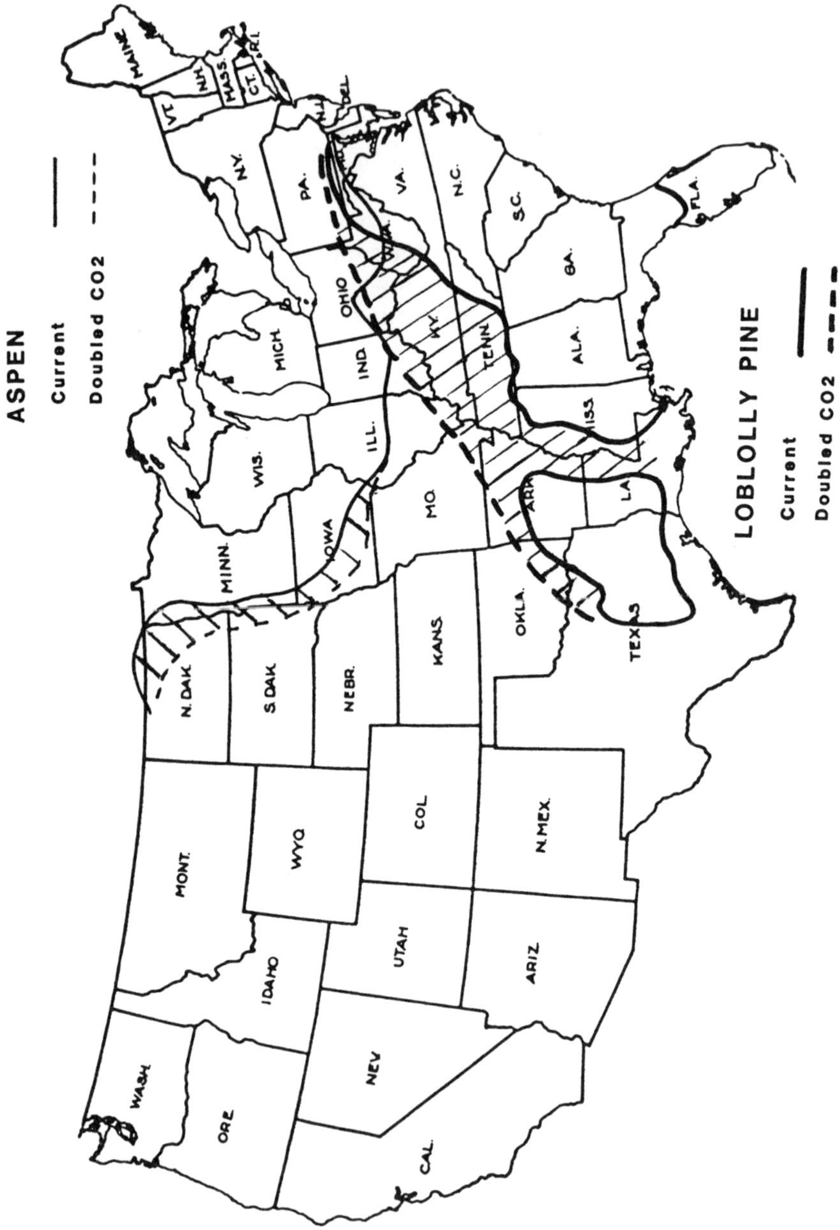

Potential range extension of aspen (westward) and loblolly pine (northward) for doubled-carbon dioxide scenario.

Table 1
Climate-Induced Impacts: Aspen-Based Industry, Great Lakes States

	Climate scenario			
	Carbon dioxide unchanged: Growth unchanged			Carbon dioxide doubled
Issue	Present	Short-term	Long-term	Probable effects / alternatives
Inventory	Surplus of over-mature age-class	Same	Balanced age-class distribution	Slightly increased growing-stock volumes
Supply	Sufficient	Sufficient	Tight	Improved
Demand	Below supply	Below supply	About equal supply	No impact
Cost of delivered wood	\$29-\$35 per cord	Stable or slight decrease	Increasing	Possibly lower
Needed utilization of merchantable timber inventory (all owners)	69%	62%	93%	Reduced
Technology	Limited use of other hardwoods	Same	Probable use of other hardwoods	No impact
Tree utilization	60%	Increased full-tree utilization	Full-tree utilization	No impact

Table 2
Climate-Induced Impacts: Loblolly-Based Industry, Southeastern United States

	Climate scenario			
	Carbon dioxide unchanged: Growth unchanged			Carbon dioxide doubled
Issue	Present	Short-term	Long-term	Probable effects / alternatives
Inventory	Unbalanced age-class distribution	No improvement	Stabilizing	Species range expanded
Supply	Few problems	Tighter	Possible	Some alleviation of supply problems
Demand	Not exceeding supply	Possible supply shortfalls	Supply shortfalls	Low impact
Regeneration	Up to $200 per acre	Increasing	Increasing	Indeterminate
Needed utilization of merchantable timber inventory (all owners)	2% of growing stock	Increasing	Close to 100%	Possible minor reduction
Technology	Few other hardwoods used	Limited substitution of other hardwoods	Increasing substitution of hardwood products	No impact
Tree utilization	70%	Increased full-tree utilization	Full-tree utilization	No impact

for loblolly pine, but the surplus of volume that existed 10 years ago has been reduced substantially. Demand for loblolly pine and other southern pines has been increasing steadily in recent years. It is expected to level off in the future especially for the higher end of the quality range, for example, peeler logs.

Climate models indicate that average winter temperatures in the southeast might increase by about 10°F or 5.5°C. Projected plant-water relations, however, remain enigmatic. Consequently, the outlook for the region's major timber species, loblolly pine, in terms of potential range shifts is favorable. Assuming the GCM scenario to be correct, there would be a potential increase in range both northward and westward (Figure 2). The extension of loblolly pine plantations into this area, however, would be severely restricted by a higher priority for agricultural land uses.

The National Forest Products Association (1980) developed estimates of supply and demand for softwoods and hardwoods for 1976 and the year 2030. Supply-demand comparisons for hardwoods and softwoods are shown in Figures 3 and 4. Impact analysis on forest industries of doubled atmospheric CO_2 concentrations should focus on those regions where current projections show a shortfall of supply by the year 2030, especially if climatic changes are expected to reduce the presence of a species, increase mortality, or reduce growth. Again assuming the GCM scenario, in the North Central region a 30% shortfall for softwoods is projected, along with a continued surplus of hardwoods. In the South Central and southeastern states, shortfalls are predicted for both hardwoods and softwoods. They are more pronounced for the South Central states with 40% and 44% shortfalls for softwoods and hardwoods, respectively. Expected shortfalls are influencing companies' current decisions on reforestation, land ownership, management programs, and research and development. However, the simulated climate scenario would probably not lead to any major change in direction of current planning.

RESEARCH DIRECTION

Because public owners do not have the flexibility of changing their location, the implications of CO_2 increases have greater potential direct importance to them and the economies within their boundaries. Government agencies should take the lead in supporting long-term research to answer the many unresolved questions and to focus on opportunities suggested by potentially broader species ranges. To carry out the types of analyses described in the framework for impact analysis, research should be expanded. Re-

Figure 3

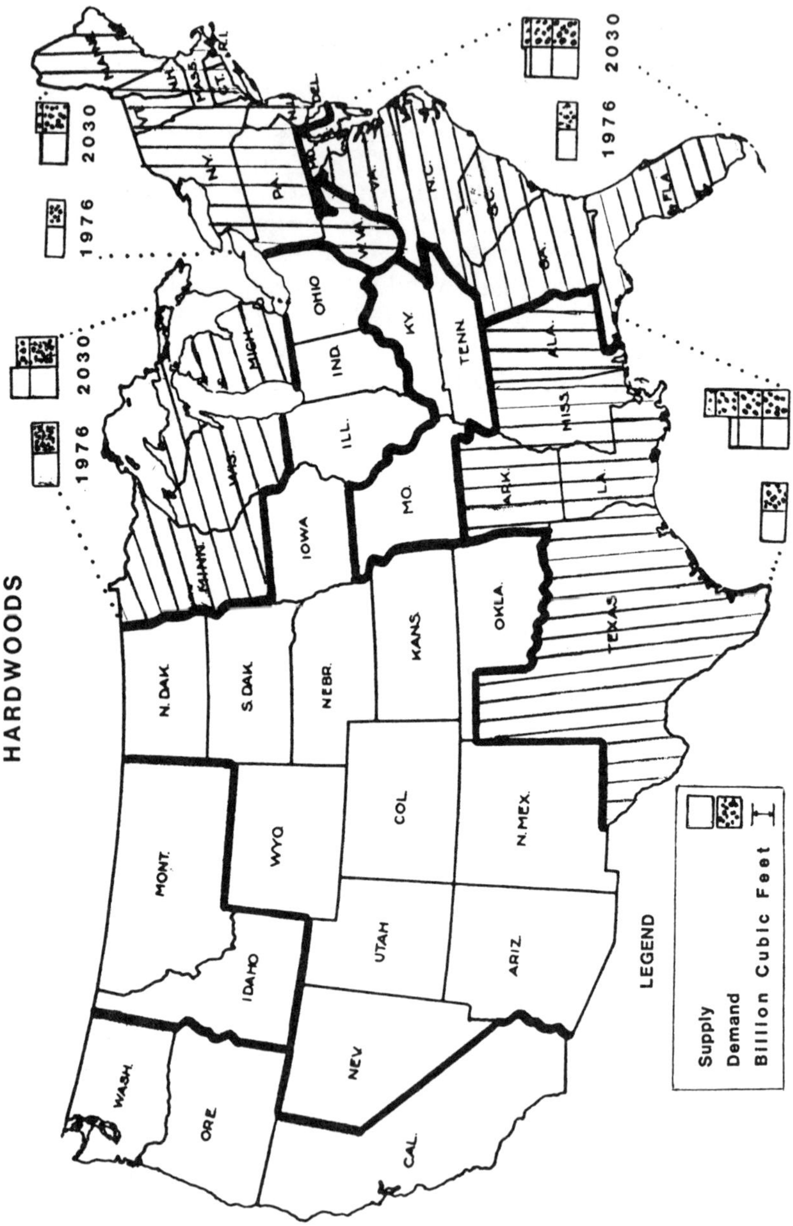

Estimated regional trends in hardwood supply and demand.

Figure 4

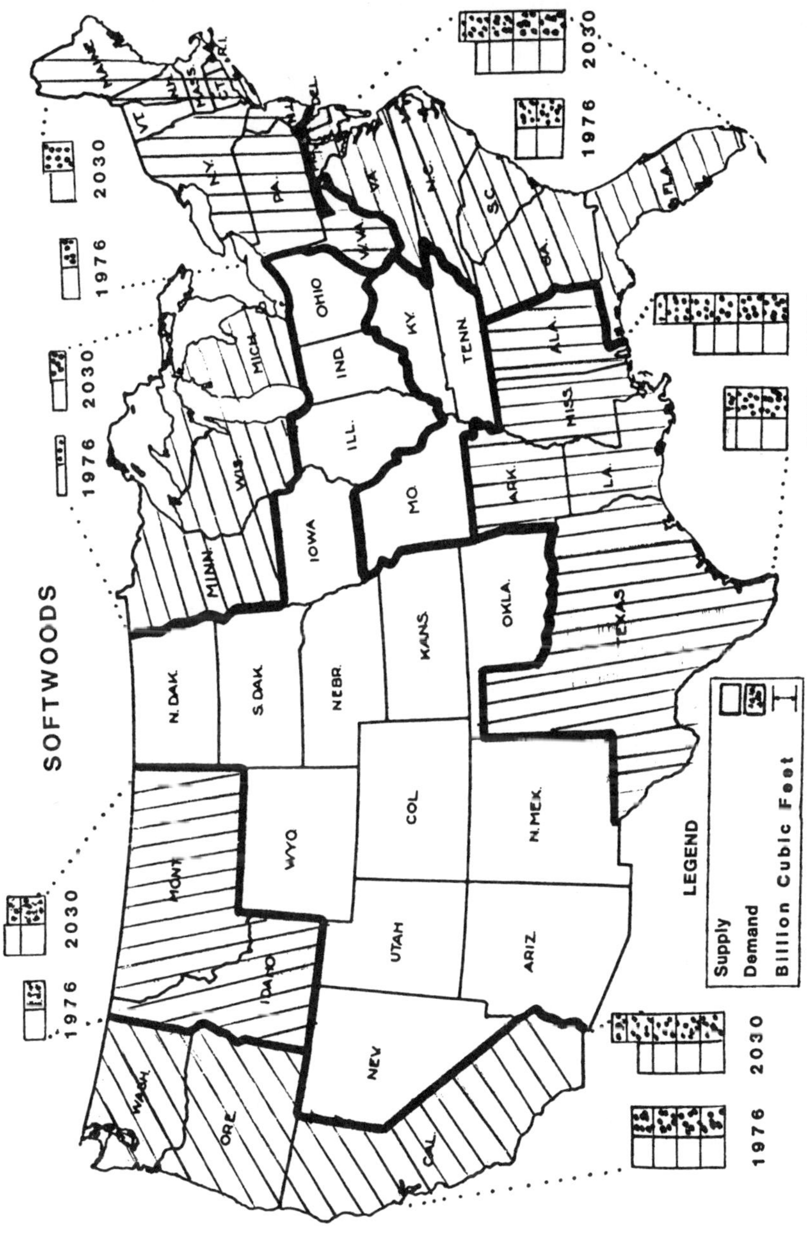

Estimated regional trends in softwood supply and demand.

search emphasis should be placed on regeneration, on refining growth models so that they incorporate climatic variables, and on estimating reduction in forestland. Related simulations using "what if" questions seem essential to developing meaningful management alternatives to possible impacts and to identifying opportunities.

ACKNOWLEDGMENTS

We gratefully acknowledge Robert N. Stone and Dwight Hair of the U.S. Forest Service for their helpful suggestions and background on regional timber supply-and-demand trends.

REFERENCES

Erkkila, D. L., D. W. Rose, and A. L. Lundgren. 1982. Estimating economic impacts of timber-based industry expansion in northeastern Minnesota. U.S. For. Serv. Gen. Tech. Rep. NC-82.

Hoffman, J. S. 1984. Carbon dioxide and future forests. *J. For.* 82, 164-167.

Montzka, T. B., and D. W. Rose. 1982. Analysis of regional timber supplies in Minnesota. Staff Paper no. 30. College of Forestry, University of Minnesota, Minneapolis.

Nadler, G. 1978. Corporate planning--a systems view. *Long Range Planning* 11, 67-78.

National Forest Products Association. 1980. Forest Industries Council Forest Productivity Report. National Forest Products Association, Washington, D.C.

O'Laughlin, J., and P. V. Ellefson. 1982. New diversified entrants among U.S. wood-based companies: a study of economic structure and corporate strategy. Univ. Minn. Agric. Exp. Station Bull. 541, Forestry Series 37. University of Minnesota, Minneapolis.

Rind, D., and S. Lebedeff. 1984. Potential climate impacts of doubled atmospheric CO_2 with emphasis on water availability and hydrology in the United States. Report prepared for Environmental Protection Agency. NASA Goddard Space Flight Center.

Rose, D. W., T. B. Montzka, D. P. Badley, and H. M. Hoganson. 1983. Assessing locational advantages of markets in terms of delivered wood costs. *Can. J. For. Res.* 13, 176-183.

Rose, D. W., A. R. Ek, H. M. Hoganson, and P. Aube. 1984. A spatial regional timber supply model for industrial expansion. *For. Prod. J.* 34, 21-27.

Smith, E. L. 1982. Evaluating development opportunities for Minnesota hardwoods. Ph.D. thesis, College of Forestry, University of

Minnesota, Minneapolis.
Williams, J. T., E. L. Smith, and D. W. Rose. 1982. Spatial distribution of Minnesota's mature hardwood inventory for 1980, 1990, 2000 and 2010. Staff Paper no. 29. College of Forestry, University of Minnesota. Minneapolis.

POTENTIAL IMPACT OF CARBON DIOXIDE-INDUCED CLIMATE CHANGES ON MANAGEMENT OF DOUGLAS-FIR AND WESTERN HEMLOCK

by James N. Woodman

ABSTRACT

No major changes in the overall management of Douglas-fir and western hemlock forests of the Pacific Northwest are foreseen as a result of climate changes attributable to doubling of atmospheric carbon dioxide (CO_2). The Douglas-fir region from the crest of the Cascade Mountains to the Pacific Ocean will have completed the transition from harvesting of old-growth natural forests to establishing and tending younger-planted forests. A trend toward prescribing management practices on a site-by-site basis should allow future managers to adjust to most local climate changes quickly and realistically. Future forest management decisions likely will be based more on economic than biological considerations. Forest regeneration may present problems in some areas, and there may be a need to develop improved seedlings and to appraise tree-thinning and fertilization practices and benefits. The most urgent research need is to develop climate models that can reliably predict seasonal climate changes for local areas within the region. The second greatest need is for yield models that can estimate future productivity of these areas, using the data from these climate forecasts.

The question, "How would a doubling of atmospheric carbon dioxide (CO_2) affect forest management in southern pine- and Douglas-fir-growing regions?" really asks, "Should we really be doing new or different forest management research over the next 10 to 20 years to prepare for this change?"

FUTURE MANAGEMENT IMPLICATIONS

If the ecological and physiological interpretations of Kramer and Sionit (this volume) and Leverenz and Lev (this volume) are correct, the future climate should be equal to or better than the present climate for the majority of sites now in the natural ranges of both western hemlock (*Tsuga heterophylla*) and Douglas-fir

(*Pseudotsuga menziesii*). Higher levels of carbon dioxide should compensate for much of the negative effects from forecast higher temperatures and lower precipitation rates. In locations where cold temperatures now limit growth, warmer temperatures will improve rates of photosynthesis and growth. Leverenz and Lev (this volume) believe that higher minimum winter temperatures may reduce regeneration and tree growth of Douglas-fir in the coastal areas of Oregon and Washington. Nonetheless, the overall outlook for both species is for an increase in their areas of natural range, increased yields, shorter rotations, and increased opportunities to manage these species profitably. Assuming the climate scenario materializes, the key assumption for this positive prediction is that the higher CO_2 levels will compensate for higher temperatures and less rainfall.

In sharp contrast to the expected adverse impact on southern pine forestry (Miller et al., this volume), few or no major changes in management of Douglas-fir and western hemlock on a region-wide basis are foreseen. If the climate forecast of Gibbs and Hoffman (this volume) becomes reality, some areas of the region will experience new drought or temperature stress-related problems, while others should develop conditions for improved forest productivity. Based on the available information received to date, the losses in productivity in one area of the region will be offset by improvements in another.

It must be emphasized that this prediction is only for hemlock and Douglas-fir growing in the west slope of the Cascade Mountains and the coastal areas. The forecast climate for the Inland Empire and Rocky Mountains regions is more adverse to tree regeneration and growth. That climate scenario would more likely affect management of major species in those regions than in the Douglas-fir region.

The doubling of CO_2 will have little effect on Douglas-fir and hemlock because, by the time of altered climate, there will be changes in management of these species but the stewardship changes will be related more to the maturing of forest management science than to climate change. The transition from an emphasis on harvesting old-growth forests to management of new second-growth forests or plantations will be completed on most commercial forestland before CO_2 doubles. Whereas previous costs of owning and managing older natural forests have been low, future costs of managing these younger forests will be much higher. It is much more difficult to justify expenditures on these new forests than it was in the past. Many companies now require scientific evidence that the dollar return from each forest man-

scientific evidence that the dollar return from each forest management practice will meet or exceed the financial returns of nonforestry investments. Many companies and public agencies are now evaluating the costs of their management practices against the expected financial benefits. This trend will become the dominant philosophy of future forest management.

FUTURE MANAGEMENT REGIMES

The greatest technological change most likely to influence future management is embodied in the concept of "site-specific management prescriptions." Although the idea of customized management prescriptions is not new, the scale of operational application so far has been relatively small in the Douglas-fir region. Forest management planning usually has focused on issues related to harvest scheduling and reforestation rather than defining a regime of management practices aimed at achievement of a stated goal. Frequently, planning and operations were oriented more to finding and applying solutions to biological problems on specific sites rather than determining whether any of these solutions made economic sense. Foresters simply were instructed to solve the biological problems at the least possible cost. Thus economic efficiency was reinforced through a performance-reward system that stressed achievement of target goals. These goals were usually based on improving average survival rates of regenerated acres, increasing average height growth of new plantations, and reducing average unit cost regardless of site difficulty or the potential financial returns from the effort required to achieve the goal. One embarrassing result of this practice was that levels of investment were not related to expected benefits. Frequently, more money was spent on low-quality land than on land with a much higher growth potential.

The recession of 1982-84 has put an end to much of this practice on industrial forestlands. Only the most productive land with favorable management costs is now likely to receive the more costly and intensive cultural treatments. If the cost of a proposed management activity cannot meet the financial criteria of the public or private landowner, there is a high probability that it will not be approved unless required by a state forestry practices law or organizational regulations.

These management adjustments relate to expected impacts from an increase in atmospheric CO_2 a couple of ways. First, this trend toward a more "customized" approach to management of Douglas-

Managers' decisions will be based on the best biological and economic information available at the time that a decision must be made. Although the combination of management practices for a particular piece of land may change over time because of new information or management objectives, the practices themselves should be the same as those used today somewhere in the region.

Second, very few managers in industry or government make major decisions based on time periods of more than 10 years. No private company lets predictions of possible events in a second rotation influence decisions in the first. Why should a company be concerned about something that may or may not happen until after most of today's owners have died? The financial climate they must deal with daily is much more volatile and more likely to affect their company's survival and health than any predicted climate change in 80 to 100 years! They must continually make major short- and long-term decisions in a changing financial climate using incomplete information about the future. Thus, it is most unlikely that a prediction of major climate change would have much influence on any major forest management decisions being made today.

Predictions of impending climate and productivity changes would most likely benefit industry managers who must make long-term strategic decisions on land acquisition or facility conversion. Should industry managers be expanding operations in one forest region or gradually moving out of another? Should they try to trade land in a future problem area for land with fewer problems and higher productivity? Will current growth and yield predictions likely be achieved within the present rotation periods? Assuming these predictions can be trusted, such information would influence these strategic planning decisions. But the relative value of such information is quite small when compared with all the other information used to make such decisions.

When would any CO_2 effects affect how forests are managed in a given area of the Douglas-fir region? If current practices are any indication, significant changes in seedling survival rates, growth rates of mature trees, or incidence of disease or insects would have to be detected first before most managers would let the changes influence their decisions. A manager's willingness to act on this new information will depend on the financial risks the organization is willing to accept.

POTENTIAL PROBLEMS IN MANAGEMENT

Although the rise of atmospheric CO_2 and related climate change would not likely affect management of Douglas-fir and western hemlock in the Douglas-fir region as a whole, managers of land in the southern mountainous and coastal border areas can expect new or more serious problems because of the "greenhouse effect."

Forest Regeneration

Without question, obtaining adequate regeneration at an affordable cost will be the number-one problem for forest managers in locations where water-stress conditions are likely to become more harsh than at present. Rates of natural regeneration and growth of western hemlock are very sensitive to high evapotranspiration conditions. Survival rates of Douglas-fir seedlings in clear-cut areas are also most sensitive to high air and soil temperatures and long periods of midsummer drought. These difficult conditions already exist in southwest Oregon and other parts of the region. Existing research indicates that it is possible to regenerate these lands but only at a high cost. Determining the trade-off between costs and expected benefits will be the major application-research question in those areas.

Expected increases in variation of seasonal precipitation rates and temperatures will cause serious problems in predicting regeneration survival rates and applying appropriate methods to ensure that those rates meet landowner requirements. Such forecasts are essential to assessing whether future regeneration efforts can justify the high front-end costs of artificial regeneration. The prospect of longer and more unpredictable cycles of drought or unfavorable temperatures increases the risk of massive regeneration failures and the possibility that such land may not be economically managed. Uncontrollable losses of major investments in artificial regeneration would make this land unattractive for private owners.

Tree Improvement

Except for the regeneration establishment problems above, few or no significant problems in the development of genetically improved seed orchards or the planting of superior-growing seedlings are anticipated. The financial returns from improved seedlings can only be realized using clear-cut harvest methods and artificial regeneration. Current methods used to select, test, and breed genetically superior trees should be adequate for projected changing climate conditions and increases in atmospheric CO_2. Realistically, genotype selections and progeny testing must be made over long periods of time under natural field conditions. It will be necessary to find the best genotypes for the new climate conditions in each breeding zone. Although the families of trees most responsive to higher levels of CO_2 could possibly be identified through enriched-growth-chamber studies, the most critical testing could not begin until the new climate conditions permitted field testing.

Some scientists believe there may be problems with these species being able to adapt to warmer seasonal temperatures. Future advances in biotechnology could offer solutions if such problems do occur. For example, a local climate's inadequate winter chilling for cone production might be corrected through genetic engineering. Likewise, inefficiencies of coastal Douglas-fir in regulating water losses might be offset by genetic engineering or by planting high-stress areas with a more drought-resistant "inland" variety of that species.

Thinning and Fertilization

The most likely problems associated with the practices of thinning and fertilization relate to predicting biological and financial benefits. Both practices require a significant commitment of dollars plus a belief that the benefits will exceed costs. Long-term estimates of growth from these treatments provide the basic justification. A rapidly changing or highly variable local climate will tend to reduce trust in existing yield models and reduce landowners' willingness to make such investments in the future.

RESEARCH NEEDS

Probably the most important question concerns research needs and priorities. What are the most important research needs of Douglas-fir and western hemlock forest managers as they relate to opportunities and problems resulting from increasing levels of atmospheric CO_2? Which studies, if any, should be initiated today to better plan for or cope with CO_2-induced climate changes?

The number-one need is to develop reliable global and regional climate models! Further models that can reliably forecast local conditions within the Douglas-fir region are needed. These models must be able to predict average seasonal patterns of precipitation and temperatures and the variation around these averages. The predictions must reflect the influences of the Pacific Ocean and the complex mountainous topography of the region. Present models do not yet have the resolution or accuracy to provide the minimum information needed. Research on global climate models needs to be accelerated so that dependent regional or local prediction models can be made.

The second most important research need is to forecast growth and yield values by using inputs from climate models and various prescribed management strategies. Such models must be able to predict regeneration survival rates for natural and artificial regeneration methods by major climate zones. They must also be able to predict differences in yields attributed to genetically improved seedlings, thinning, and fertilizations. Obviously this research cannot be started until better climate models are available and the climate has changed sufficiently to allow trees to respond to the new conditions.

Finally, long-term field studies are needed to inform forest managers about the costs and benefits of all management practices in different local climates and at higher atmospheric CO_2 concentrations. Obviously, knowledge about future climate conditions is too uncertain at this point to determine when and where such field testing should be done. Ideally, the field testing should be on sites where the soils and climate conditions will be similar to future growing conditions. It will be at least 50 years before enough of these conditions exist to begin such studies. The previously described trend toward more site-specific management should indicate both where and when such studies are appropriate.

BREEDING STRATEGIES IN FOREST TREE POPULATIONS TO BUFFER AGAINST ELEVATED ATMOSPHERIC CARBON DIOXIDE LEVELS

by R.C. Kellison and R.J. Weir

ABSTRACT

The goal of tree improvement programs is to produce genetically improved plant material to establish plantations. The accepted procedure is to select trees of good growth, form, and pest resistance from natural stands and plantations for inclusion in vegetatively propagated seed orchards. By progeny testing, the poor genotypes of the selected trees are identified and rogued from the orchard, leaving only the best trees to cross-pollinate. The second purpose of the progeny test is as a base population for the second-generation clonal orchard, its progeny serves as the base for the third generation, and so on.

A concern exists that the geneticist cannot react quickly enough to avert disaster in a tree-breeding program from elevated levels of carbon dioxide (CO_2). The concern results from the long time required for forest trees to attain sexual and economic maturity. The opposing reaction from the tree breeder is that forest trees are inherently heterozygous at the genetic level because of the lack of artificially directed selection, and that the effort is to maintain and even expand genetic diversity. Because of the relatedness among trees within natural stands and among trees established from seeds of unimproved sources (where the seeds are customarily collected from one or a few trees), the genetic diversity is greater within a stand of trees established from genetically improved seed than it is in ex situ stands.

The genetic diversity incorporated in genetically well-buffered tree populations could theoretically result in superior performance from elevated atmospheric CO_2 levels. However, greenhouse tests from which such results have been obtained have held all other variables constant while CO_2 levels were varied. Despite the best plans of the geneticist, uncertainty will exist about the effects of elevated CO_2 levels on plant life until the interacting effects of the genotype and the environment can be evaluated.

Tree improvement programs in the United States, especially in the South, have proliferated since the end of World War II. By 1981

13,600 acres of forest seed orchards had been established (U.S. Forest Service, 1982), with the goal of using genetically improved plant material from these and additional orchards on every artificially regenerated acre. The members of the North Carolina State University-Industry Cooperative Tree Improvement Program own or control about 20 million acres of forestland and plant over 700,000 acres annually, 80% of which is done with genetically improved plant material (North Carolina State University, 1984), and the goal of 100% self-sufficiency is obtainable within five years. Other organizations in the South and in other areas of the United States are striving to fulfill similar objectives.

The question being asked is whether the selection and breeding strategies of today are adequate to maintain long-term productivity, and even to avert catastrophe, in an environment with atmospheric carbon dioxide (CO_2) levels projected to increase substantially within the next 100 years (Abelson, 1983). Experiments with tree improvement show that well-designed programs have the flexibility to react to changing environmental conditions and that breeding strategies for elevated amospheric carbon dioxide levels will be no different from those used for other types of potential gains and losses.

PROCEDURES OF CURRENT TREE IMPROVEMENT PROGRAMS

Most tree improvement programs in the United States and elsewhere in the world are based on mass selection. The procedure is to determine the species having greatest local importance and, within those species, to identify the kind and amount of variation. If a species has clinal variation, such as occurs with loblolly pine (*Pinus taeda*), in which a trait varies in a predictable and genetically stable fashion with changes in latitude, elevation, and precipitation, artificial breeding zones are established along political boundaries or natural landscape features. Conversely, if ecotypic variation occurs, such as has been shown for yellow-poplar (*Liriodendron tulipifera*) (Kellison, 1970), in which certain tree traits are unique to certain populations, rather exact breeding zones are established, regardless of political boundaries and natural geographical features. Experience shows that plant material from species having clinal variation can be moved limited distances outside of the artificially imposed breeding zone without significantly changing the tree's performance. The opposite is true for species having ecotypic variation; the best tree performance occurs when the progeny are planted on the land type from which the parents originated.

The type of tree variation within the breeding zone of either a cline or an ecotype must also be determined. Genetic tests show tree populations of indigenous species to be significantly different from one natural stand to another when seed from the individual trees are bulked. When the seed lots are kept separate by individual tree, however, the genetic variation among trees is so great that the stand differences are masked. For example, the specific gravity of wood among trees within a natural stand of loblolly pine at 40 years of age within the lower coastal plain of North Carolina will range from about 0.44 to 0.64 (mean = 0.54 unextracted) (Zobel et al., 1972). The mean difference among stands within the province would range only from about 0.54 to 0.56. For a trait such as wood specific gravity, which is under strong genetic control in the southern yellow pines and many other species, greater variation exists among trees within a stand than among stands, among geographical areas, and even among species. The greatest genetic gains can be obtained from selecting the best trees of the best stands of the best species--mass selection.

The phenotypically select trees included in most tree improvement programs have been chosen because of their superior attributes in growth, form, adaptability, freedom from pests, and desired wood properties. The most commonly accepted practice thereafter is to incorporate the selected phenotypes in a clonal seed orchard, established by grafting, budding, and rooting, before the genetic worth of the parent trees (clones) is determined. The inferior clones, the genotypes of which are determined from progeny tests, subsequently are rogued from the orchard, and the remaining best trees are managed for optimal seed production, from which operational plantations are established.

In addition to providing information for determining the genetic worth of the clones, progeny tests of the first-generation trees serve as the source of material for the second-generation seed orchard, their progeny are the base for the third generation, and so on. Special precautions are needed to avert inbreeding depression that can result from recombination and continued selection in a restricted genetic base.

GENETICALLY IMPROVED PLANT MATERIAL

The purpose of tree improvement programs is to produce genetically improved plant material for the establishment of plantations. Because of logistics, only 25 to 30 select trees are included in a typical clonal seed orchard. Each of those trees is grafted, or otherwise vegetatively propagated, an equal number of times to build a seed orchard of desired size. A purported limitation to this approach is that the genetic base is restricted over that obtained in a natural stand and over that existing in a plantation established from wild seed. For the southern pines, and likely for many other species, that contention is false. The practice observed worldwide for plantations established from wild seeds is that the seeds are collected from one or a few trees in one or a few stands. The added limitation is that the preponderance of seed comes from trees of large crown and poor form because of the ease of seed collection.

The opposite situation occurs in a tree improvement program: only one tree from each of 25 or 30 stands is included in the seed orchard. Logic dictates that best performance would be obtained from random pollination among the 25 or 30 selected clones, as opposed to a mix of seeds from several unselected trees in the same stand. Progeny test results show the logic to be correct. And an unexpected gain is obtained. The performance of either half-siblings or full-siblings--that is, either only the mother is known (half-siblings) or both parent trees are known (full-siblings)--from seed orchards of southern pine is consistently better in growth and volume production than the seed production area's check lot, which originates from a mixture of seeds from the best trees of a single stand. The apparent reason for the inferiority of the seed production area's material of the southern pines is that only a few trees remained in fencerows and along watercourses to seed in fields abandoned after soils were depleted from intensive agricultural cropping at the turn of the century. High levels of coancestry, related mating, and subsequent inbreeding depression are apparently occurring within a stand, and outcrossing occurs when single trees from separate stands are allowed to cross-pollinate in a seed orchard. The added gain obtained in growth and volume production, but not in tree form and resistance to fusiform rust (*Cronartium fusiforme* f. sp. *fusiforme*), supports this supposition. Inbreeding depression is noted for adversely affecting survival and growth traits more than form and quality traits.

The greatest genetic gain from a tree improvement program is obtained when only the genetically best clones in a seed orchard are used. Many orchards in the North Carolina State University-Industry Cooperative Tree Improvement Program have been rogued to leave only 12 to 15 of the best clones from the original 25 to 30. Within that select list, some organizations collect by parent group (best, average, or poorest, for example), although cross-pollination ostensibly occurs among all 12 to 15 clones. The seed and seedling mixture from the five or six best clones, for example, creates more uniformity than the progressively larger number of clones, but it is still relatively broad compared to plantings made from the seed and seedlings of single-clone collections, and the single-clone collection is an order of magnitude broader in genetic diversity than a single genotype propagated vegetatively.

All of these options are being practiced to a degree with forest trees. Soon tissue-cultured plantlets will be available for all forest trees, allowing the establishment of vegetatively propagated forests such as is now only practiced with selected species of *Eucalyptus*, *Cryptomeria*, *Populus*, *Salix*, and *Picea*.

MAINTAINING GENETIC DIVERSITY

Compared to field crops, even short-rotation forestry of 5 to 10 years is long term. An additional risk is encountered in forestry for every year the rotation is extended and for every portion of each year after agricultural crop harvest that the tree crop remains exposed to the elements.

A risk in a tree improvement program, other than catastrophic events, is the uncertainty about the direction and magnitude of selection over a long time span. Some geneticists have refrained from including a trait, such as specific gravity, in an otherwise operational breeding program because of the concern that the type of wood desired 50 years hence is unknown. Others have initiated tree improvement programs with low-to-moderate selection intensity because of a fear of changing the trait so drastically that the resource would be unusable at rotation age (Silen, 1966).

Both tacks seem to be half-hearted attempts. In the first instance, forest trees, which have occupied a position on earth for thousands of years and which have been without directed selection until about 30 years ago, are extremely heterozygous and therefore genetically diverse. In comparison, the genetic diversity of corn (*Zea mays*), which has been selectively bred annually for hundreds of years, continues to exhibit tremendous variability. To think that the genome of loblolly pine or Douglas-fir (*Pseudotsuga menziesii*) could be changed so drastically in one generation of breeding borders on the ludicrous. It seems equally as farfetched to suggest that offspring from one generation of breeding will be significantly more prone to pest attack than are unimproved offspring. Nevertheless, in a plant-breeding program of any sort, and especially a forest tree-breeding program, precautions must be taken to avert potential disaster, in both the short and the long runs. The best insurance is to maintain a broad genetic base from the start.

The Cooperative Effort

The North Carolina State Tree Improvement Program consists of 40 operating units bounded to the west by the Mississippi River and to the north by the Mason-Dixon Line. Each operating unit has at least one loblolly pine improvement program consisting of an orchard with 25 or 30 clones; some organizations have two or three first-generation loblolly pine seed orchards at the central location. More than 2300 trees, all of which have been standardly graded, have been accepted into the program for seed orchard use and progeny testing. Results from the progeny tests are being used to rogue the first-generation orchards, and the superior trees of the superior families are being selected from the progeny tests for establishment of the second-generation orchards. More than 500 acres of second-generation orchards have been established since their beginning in 1974.

Supplementing the Genetic Base

The need to supplement the genetic base with unrelated clones from natural stands and from plantations established from unimproved seed was recognized a few years ago. More than 3100 new loblolly pine selections were added to the base. A six-clone disconnected-diallel mating design is being used to test the progeny of these selections. The superior phenotypes will be inserted into the mainline program for establishment of the third-generation seed orchards.

Dual-Phase Tree Improvement

The principal emphasis in tree improvement programs has been in the production phase--that is, the production of genetically improved seed for the establishment of operational plantations. Of greater importance in the long run, however, is the need to maintain a breeding phase. The purpose of the breeding phase is to continually strive to create populations of new genotypes that will allow the succeeding-generation orchard to be superior to the present one and to maintain a broad genetic base in perpetuity. The latter objective is even more important than the former because of a moral obligation to help avert genetic catastrophes in the future forests.

THE GREENHOUSE EFFECT ON FOREST TREE POPULATIONS

At the 1973 rate of fossil fuel combustion, atmospheric carbon dioxide levels are projected to double in 57 years, raising global temperatures an estimated 3°C (5.4°F). During the past decade, however, the burning of fossil fuels has stabilized, increasing the projected time for doubling of atmospheric carbon dioxide to 220 years (Abelson, 1983). The "greenhouse effect," resulting from the doubling of atmospheric carbon dioxide, has been heralded, by some prognosticators, as disastrous to plant life. However, another view holds that genetic diversity within a wide-ranging tree species, such as loblolly pine, is sufficiently broad to allow for directed management in new environments that fall short of catastrophic levels. The astute manager may even realize benefits.

Greenhouse tests with various forest tree species, including eastern cottonwood (*Populus deltoides*) (MacDonald, 1982) and ponderosa pine (*Pinus ponderosa*), Scotch pine (*Pinus sylvestris*), and blue spruce (*Picea pungens*) (Lemon, 1976), show photosynthesis and short-term growth rates doubling at carbon dioxide levels two to five times normal. Similar results have been obtained from experiments with loblolly pine and sweetgum (*Liquidambar styraciflua*). However, most controlled experiments show photosynthesis to culminate at carbon dioxide levels about two times normal, the point at which water transpiration decreases about 25% (MacDonald, 1982).

These results suggest that tree growth and fruit production would increase as atmospheric carbon dioxide does; however, the greenhouse experiments are invariably conducted in the presence of ample light, moisture, and nutrients. Because elevated atmospheric carbon dioxide levels would increase global temperatures, water levels, and soil acidity and would decrease the soil nutrients essential for plant growth as photosynthesis was enhanced, it is difficult to determine the effect on plant survival and growth. Evidence suggests that control-pollinated families of Douglas-fir (Lambeth, 1979) and loblolly pine (Ledig, 1967) react differently to varying amounts of light, moisture, and nutrients in a greenhouse setting; indeed, soil moisture, in these experiments, accounts for more of the genetic variation than any other variable. The genetic correlation between 12 greenhouse environments and field performance of the 16 Douglas-fir families was only 0.39, indicating the fallibility of testing in a controlled environment for an uncontrolled environment.

The evidence is overwhelming that elevated levels of carbon dioxide can affect a plant's photosynthetic and respiration rate, but the evidence is less clear and even lacking that the gas will affect species composition. One valid speculation is that species composition would first be affected in extreme environments, such as those that support high-altitude forests. The recent decline in red spruce (*Picea rubra*) forests of the southern Appalachian Mountains could be the result as much of atmospheric carbon dioxide buildup as it is of atmospheric ozone accumulations and acid deposition.

The irony of this situation, as obvious as the possibility might seem, is that scientists are able to effectively study only one potential cause at a time (Postel, 1984). The study of interacting influences is so complex and the results so open to interpretation that the topic is studiously avoided. In the meantime the world's plant life is absorbing some abusive shocks. The animal life has no recourse but to follow!

SUMMARY AND CONCLUSIONS

Despite the exhibited impacts on certain of the world's forests, tree species such as loblolly pine, Douglas-fir, and sweetgum, with their broadly adapted genetic bases, will be among the last forest trees to be adversely affected by rising CO_2 levels. Well-designed tree improvement programs, in which there is a conscious effort to maintain genetic diversity by cross-breeding and genetic engineering, will enhance the ability of a species to survive under progressively declining conditions. After all, the best-adapted trees are selected at each generation. The shorter the generation interval, the more likely that managed tree populations will rapidly respond through selection to changing environmental pressures.

The argument has been made that gene conservation is essential for species used in intensive tree improvement programs. The justification is that genes associated with economically unimportant traits in current tree improvement programs, such as slow tree growth, might have redeeming value in future generations for pest resistance, for adaptability of the species to a changed environment, or for new tree products. The most often recommended procedure is to preserve populations of plants (trees) ex situ. But it is equally likely that ex situ populations of commercially important tree species would be especially vulnerable to large environmental changes. Most tree species do not immediately succeed themselves under the best circumstances. A likely adaptation to prevent site degradation and reestablishment of the population under a significantly changed environment would be impractical if not impossible. New gene combinations appear to be the order of a changing universe, whether change is natural or the result of human perturbations. A well-designed and carefully implemented tree improvement program offers the greatest opportunity for a future day.

REFERENCES

Abelson, P. H. 1983. Carbon dioxide emissions. *Science* 222, 889.

Kellison, R. C. 1970. Phenotypic and genotypic variation of yellow-poplar. Ph.D. Thesis. School of Forest Resources, North Carolina State University, Raleigh.

Lambeth, C. C. 1979. Interaction of Douglas fir (*Pseudotsuga menziesii* (Mirb.) Franco) full-sib families with field and phytotron environments. Ph.D. Thesis. School of Forest Resources, North Carolina State University, Raleigh.

Ledig, F. T. 1967. Variation in photosynthesis and respiration in loblolly pine (*Pinus taeda* L.) progenies. Ph.D. Thesis. School of Forest Resources, North Carolina State University, Raleigh.

Lemon, E. 1976. The land's response to more carbon dioxide. Paper presented at the U.S. Office of Naval Research Conf., "The Fate of Fossil Fuel CO_2." Honolulu, Hawaii.

MacDonald, G. J. 1982. *The Long-term Impacts of Increasing Atmospheric Carbon Dioxide Levels.* Ballinger Publishing Co., Cambridge, Mass.

North Carolina State University. 1984. Twenty-Eighth Annual Report: N.C. State University-Industry Cooperative Tree Improvement. School of Forest Resources, North Carolina State University, Raleigh.

Postel, S. 1984. Air pollution, acid rain, and the future of forests. Worldwatch Paper 58. Worldwatch Institute, Washington, D.C.

Silen, R. R. 1966. A simple, progressive tree improvement program for Douglas fir. PNW-45, USDA Forest Service, Portland, Oreg.

U.S. Forest Service. 1982. *1981 Directory of Forest Tree Seed Orchards in the United States.* USDA Forest Service FS-278. U.S. Government Printing Office, Washington, D.C.

Zobel, B. J., R. C. Kellison, M. F. Matthias, and A.V. Hatcher. 1972. Wood density of the southern pines. Tech. Bull. No. 208. North Carolina Agricultural Experiment Station, North Carolina State University, Raleigh.

FORESTRY RESEARCH NEEDS AND STRATEGIES

by J. Charles Lee and Paul J. Kramer

ABSTRACT

An improved understanding of how rising atmospheric carbon dioxide (CO_2) affects forests could provide technological information to temper ad hoc political decisions. However, problems of rising CO_2 and climate change rank low on the forestry research agenda. One reason is the breadth and magnitude of the problem; no single agency bears lead responsibility. Further, forest scientists have difficulty dealing with complex problems affecting long-term forest productivity--as with acid precipitation. Finally, industry's economic planning horizons are too short to generate actions with uncertain, distant benefits. The question is not one of whether scientists can provide conclusive proof of a problem, but whether society will take action necessary in anticipation of future change.

Most of the gaps in knowledge are amenable to conventional research methodology. Regardless of the potential effects of rising CO_2, there is much we need to know about tree physiology and tree responses to environmental and biological stress. A strategic approach to research would focus on direct versus indirect effects of CO_2 buildup.

While we do not know for certain how the the increase in atmospheric CO_2 will affect global climate, we are beginning to understand how increased concentrations of CO_2 affect trees under laboratory conditions. It would be useful for scientists to improve their understanding of how present weather extremes affect the physiological processes which control tree growth. As scientists learn more about the physiological bases of drought and temperature tolerance, productivity should improve in the short term and we would be better prepared to deal with future climate changes, whatever they may be.

For the future, forest scientists should become more effective in competing for research funds in basic ecological and biological research. Further, forestry should integrate its knowledge base with that of other plant and ecological systems through collaborative, multidisciplinary efforts.

While there is broad agreement that carbon dioxide (CO_2) levels will continue to increase, the uncertainties about its effect on climate, tree growth, and forest ecosystems pose difficult choices for forest managers, forest scientists, and government agencies. While some effects will be beneficial and likely offset some negative impacts, how they will balance out is unknown. By anticipating change, the forestry community can avoid surprises, identify alternatives, and seek to temper ad hoc political decisions with technological information.

Accelerating research is a logical response to uncertainty about the future. Research can provide the knowledge needed to maintain adaptability in forest management systems so they can adjust to rising levels of atmospheric CO_2 and whatever changes in climate may occur. Fortunately, forestry is more future-oriented than most economic endeavors, and many forest managers and scientists recognize the need for additional information on which to base future changes in management strategies.

Uncertainty and risk are unavoidable in forest management, and forest managers are accustomed to dealing with them. They must assess, for example, the likelihood that a tree planted today will not live to maturity, what impact new forest products technology will have on the value of the tree when it is harvested, what the balance between timber supply and consumer demand is likely to be 40 or 50 years in the future, and whether changes in public attitudes about forest practices are likely to constrain management or require more costly management practices. While strategic planning and future forecasting are receving greater attention in corporate boardrooms and forestry agency offices, problems of CO_2 and climate change appear to rank low on the lengthy list of uncertainties faced by investors, managers, and public administrators. The reasons for this are several.

ECONOMIC AND SCIENTIFIC BARRIERS

The very breadth of the problem and the magnitude of its consequences confound attempts to address issues related to the buildup of CO_2 and other "greenhouse" gases in the atmosphere and consequent climate change. Since the generation of CO_2 and its consequences are not constrained by state boundaries and are, in fact, global in scope, responsibility for coordinating a national response logically lies with the federal government. But the problem also spans the missions and responsibilities of a number of federal agencies--the National Oceanic and Atmospheric Administration, the USDA Forest Service, the Department of Energy, and the Environmental Protection Agency, to name but a few. Who, among the agencies, should take the lead in research on greenhouse gases, climate change, and tree and forest responses? This diffusion of responsibility inhibits development of a coordinated research program.

Forest scientists have difficulty answering fundamental questions about complex problems affecting the long-term productivity of forests. How acid precipitation will affect forests is a good example. Recently, an assistant to the governor of a southern state complained that foresters always give weak answers when asked to explain why the rate of growth of several economically important species apparently is declining. A scientist responded that uncertain answers resulted from inadequate knowledge. While this situation can be blamed in part on anemic levels of support for forestry research, it also reflects the difficulty facing forest scientists when they are faced with complex issues which require the concentrated attention of many disciplines. More interdisciplinary integration is necessary to ensure that forestry benefits from knowledge developed elsewhere.

Forest technologists frequently find themselves applying techniques developed in other biological fields. For example, biological technology often is applied to field crops long before it can economically be applied to trees. While this sometimes is cost effective, it does not encourage the development of the broad scientific capability in forestry required for an understanding of the principles and processes of the technology being adopted. This lack of capability puts forest managers at a disadvantage when dealing with issues that are peculiar to forestry.

Economic viability is the standard against which industry investments and many government programs are judged. Investment planning horizons usually are too short to deal effectively with issues--like many relevant to forestry--where action is required today, but the payoff may not be apparent for decades. Neither taxpayers nor corporate stockholders are notably tolerant of investments directed at uncertain problems. Developing a consensus is difficult even when the problem is immediate and recognized; the difficulty increases by several orders of magnitude when the health, safety, or pocketbook effects of a problem are subtle and will not be obvious--especially to the layman--for many years. Lastly, some effects of CO_2 enrichment will be positive, at least within limits.

RESEARCH NEEDS

All of these factors contribute to the lack of a sense of urgency about CO_2 and climate change and raise obstacles to accelerating research. However, neither complexity nor uncertainty is an acceptable reason for maintaining the status quo, nor is the fact that scientists cannot agree on the potential seriousness of the problem or options for dealing with it. The question is not whether scientists will be able to provide conclusive proof that there is a risk in the future because of rising CO_2 and its potential effects on climate and resources on which society depends, but rather whether society can and will take action necessary today in anticipation of future change.

The scientific basis for forestry must be strengthened. Typically, agriculture and forestry research scientists have emphasized short-term applied research, in part because of the time horizons dictated by the availability of research money. Over the past five years, however, there has been an increased recognition of the need for targeted, mission-oriented research on basic principles and processes. The U.S. Forest Service and the Association of State Colleges and University Research Organizations (ASCUFRO) have published an inventory of basic research needs which includes research on CO_2.

Most of the known gaps in knowledge are amenable to conventional research methodology. Moreover, as atmospheric and forest scientists fill in the knowledge gaps on topics bearing on accumulation of greenhouse gases, climate effects, and tree responses, they also will contribute to a greater understanding of weather and climate, tree and stand response to environmental and biological stresses, and ways of mitigating stress through breeding or modification of the environment of individual stands. Regardless of the effects of CO_2, we need to know more about tree physiology in order to guide breeding, protection, and cultural activities--particularly with the advent of promising new biotechnologies. This research should not be charged solely to concerns about CO_2-induced change, but to the need for a better understanding of biological, ecological, and economic factors affecting the viability of the forest resource.

The research agenda is long, and many of its elements are readily apparent. In the area of atmospheric and climate change, a better understanding of the global carbon cycle is necessary to improve predictions of future CO_2 concentrations and the relationship between accumulations of atmospheric CO_2 and climate change. Global climate models which incorporate cloud and ocean influences and better describe regional and seasonal conditions are needed to predict regional and local effects. Better models also would help distinguish the impacts of weather from those of long-term climate change.

In terms of biological processes, a better understanding is needed of how important tree species respond in growth and development to higher levels of ambient CO_2 over a range of temperature, light, and moisture conditions, and how important species respond to enriched CO_2 environments over the long term and in the presence of environmental stress. There is a need for a better understanding of seedling and site characteristics which influence stocking and growth on stressed sites and of the effects of CO_2-induced change on tree-pest relationships. Models which can explain and predict CO_2-induced effects on species relationships and structure and function in ecosystems, both managed and unmanaged, will help the forest manager predict future impacts on trees being planted today. We also need to know more about possible interactions between increasing CO_2 concentration and air pollutants with respect to tree growth.

Armed with information on anticipated climate change and likely tree performance under different situations, the forest manager can incorporate these considerations into an overall management strategy. An array of options might be available: to acquire land more favorable for growing trees in the new environment, to modify regeneration systems, to alter breeding strategies or species, cultural practices, or product mix. These options might capture the fertilization effects of higher atmospheric CO_2 levels, while mitigating the effects of warmer, drier environments.

A STRATEGIC APPROACH TO RESEARCH

It is probable that tree growth already is being affected by increasing concentrations of CO_2 in the atmosphere and that these effects will become increasingly important during the next two or three decades. Questions concerning the direct effects of CO_2 are important today, and available resources can be focused on specific elements where base information already is available. As a strategy for researching the possible effects of CO_2 on forest environments is developed, it should be kept in mind that it is not necessary that questions concerning future climatic conditions be resolved before useful research is begun on the effects of environmental stresses interacting with CO_2. Research focused on how water, temperature, and other environmental stresses affect tree growth today can be combined with studies of climate change to give us some idea of how tree growth will change in the climate of the next century or so.

Direct versus Indirect Effects of CO_2 Buildup

When developing a research strategy, it is useful to distinguish between the direct or fertilizer effects of increased CO_2 on plant growth and the indirect effects of changes in climate. While scientists still debate the "greenhouse effects" of gases on the global climate, the direct effects of increased concentrations of atmospheric CO_2 on trees are beginning to be well documented under laboratory conditions.

It is well established that increasing the the CO_2 concentration above the present ambient concentration results in increases in height, dry weight, and often in stem diameter and leaf area of tree seedlings (Kramer and Sionit, this volume). Exposing seedlings to increased CO_2 also causes partial closure of stomata, reduction in transpiration, and increase in the rate of photosynthesis, resulting in a significant increase in water-use efficiency. While the direct or fertilizer effects on first-year seedlings are well established, there are no data as yet to indicate whether the increase in growth continues into the sapling stage.

However, the warming and drying postulated for parts of the United States and Canada under the doubled-CO_2 scenario might at least partly override the fertilizing effects of increased CO_2. Experiments on a few species indicate that the growth response to CO_2 is considerably reduced by environmental stresses such as droughts and unfavorable light and temperature conditions. Thus the degree of a species response to increasing CO_2 will depend on weather and climatic conditions, particularly changes in rainfall and temperature. When a plant fails to receive sufficient water, cell growth slows and photosynthesis and other metabolic processes are affected. However, a high concentration of CO_2 will partly compensate for water stress and high temperatures. Thus the interaction between CO_2 and environmental factors that control growth deserves intensive investigation.

Variations in Weather versus Climate Change

Climatologists agree that there will be a warming trend with a global average increase in temperature of 2 or 3°C (4 to 5°F), but with a larger increase in the Arctic. Global patterns of rainfall also will change, with some regions becoming warmer and drier while other areas become moister. Unfortunately, present models cannot accurately predict changes for specific regions.

Weather records, however, indicate that there have been year-to-year variations in temperature and rainfall much greater than the overall increases expected because of the greenhouse effect. It is reasonable to surmise that for several decades at least relatively short-term variations in temperature and participation will have more influence on tree growth than the gradual climate changes resulting from increases in greenhouse gases in the atmosphere.

Until meteorologists develop the methods and models to more accurately predict how the climate will change globally and regionally, it would appear more productive for plant physiologists and forest scientists to concentrate on improving their understanding of how current weather extremes affect the physiological processes that control tree growth. Since drought is the most common cause of reduced tree growth, it is especially important to see if tolerance to drought can be improved. If scientists learn more about the physiological bases of drought and temperature tolerance, productivity could be improved in the immediate future and we will be prepared to deal with future changes in climate, whatever they prove to be.

Further, many foresters today are managing sites which have the temperature and moisture characteristics that might occur on a broader scale under a CO_2-enriched scenario. Techniques for managing these stressed sites should be continually researched and refined. If we are successful in understanding how variations in weather affect tree growth and how we can increase tolerance of weather stress, that we will have the information needed to cope with a changing climate.

THE FUTURE

There is a clear need to encourage more research to improve our understanding of how greenhouse gases affect climate and to sharpen the ability of atmospheric scientists to predict what will occur. The forestry community needs a better understanding of how weather and climate affect the forest environment generally. Forest scientists are well along in understanding the direct effects of increased CO_2 on tree and plant growth in the short term, but a great deal of additional work is required before we understand the cumulative effects of increased atmospheric CO_2 and climate change.

It is clear that forest scientists will have to become more effective in competing for funds for basic ecological and biological research provided by such agencies as the National Science Foundation, the Department of Energy, and the Department of Agriculture. This will require some new approaches. A modest campaign will have to be mounted to make these agencies aware of basic research needs peculiar to long-lived species, the urgency of those needs, and their relationship to the nation's social and economic welfare. Forest scientists will have to be able to prepare proposals in terms consistent with each agency's mission and project requirements. It also is important that scientists involved in peer review of forestry research proposals understand and take account of difficulties inherent in using trees as experimental organisms. While much of the fundamental knowledge gained from studies with other plants will be transferable to trees, this will not eliminate the need for a sound forest research effort; studies peculiar to perennial species will still depend largely on the traditional forest scientists.

There is a need for more integration of the knowledge base in forestry with that of other plants and ecological systems. Forestry should position itself to take advantage of knowledge developed in other disciplines. Incentives should be developed to encourage collaborative multidisciplinary efforts. Forest scientists should advocate a greater effort by federal agencies which have a responsibility for atmospheric research and should stay abreast of this work. Technical journals in forestry should solicit manuscripts from atmospheric scientists so that forest scientists will be aware of likely future events of significance to tree growth and forest productivity. Interdisciplinary symposia should be held periodically to provide forums for exchanging knowledge and for refining the research agenda--particularly in those areas which would benefit from interdisciplinary collaboration. For its part, the forest science community should consider greenhouse gas-induced climate change scenarios when setting research priorities and designing research projects.

If the forestry research community can overcome obstacles, seize opportunities, and confront challenges, we might see the development of a research capability adequate to provide the knowledge base necessary to resolve issues like acid precipitation, CO_2-induced climate change, and biotechnology in a timely manner. Future survival may well depend on ecological systems and management programs sufficiently elastic to sustain moderate failures in one or several parts without causing catastrophe. By anticipating change and generating the knowledge to adjust to it, foresters can ensure the sustained productivity of the nation's and world's forest resource to meet human needs.

ABOUT THE AUTHORS

Philip M. Dougherty is Assistant Professor of Forest Resources at the University of Georgia, Athens, Georgia.

Inez Fung is a Physical Scientist at the NASA Goddard Institute for Space Studies, New York City.

Michael J. Gibbs is Project Manager with ICF Incorporated, Washington, D.C.

Dr. James Hansen is Head of the NASA Goddard Institute for Space Studies, New York City.

John S. Hoffman is Director of the Strategic Studies Staff, U.S. Environmental Protection Agency, Washington, D.C.

Paul J. Kramer is the James B. Duke Professor of Botany Emeritus at Duke University, Durham, North Carolina.

Andrew Lacis is a Physical Scientist at the NASA Goddard Institute for Space Studies, New York City.

Sergej Lebedeff is a Scientific Analyst for SIGMA DATA MA/Com Corp.

Deborah J. Lev just graduated at the College of Forest Resources, University of Washington, Seattle, Washington.

Jerry W. Leverenz is a Research Associate at the Department of Plant Physiology, University of Umea, Umea, Sweden.

Frank W. Miller is Professor of Forestry and Director of the Remote Sensing Center at Mississippi State University, Mississippi State, Mississippi.

David Rind is a Physical Scientist at the NASA Goddard Institute for Space Studies, New York City.

Dr. Walter Orr Roberts is President Emeritus of the University Corporation for Atmospheric Research, Boulder, Colorado.

Gary Russel is a Mathematician at the NASA Goddard Institute for Space Studies, New York City.

William E. Shands is a Senior Associate with The Conservation Foundation, Washington, D.C.

Nasser Sionit is a Research Professor at San Diego State University, San Diego, California.

George L. Switzer is Professor of Forestry at Mississippi State University, Mississippi State, Mississippi.

John Bruce Wells is President of Bruce Company, Washington, D.C.

James N. Woodman is Director of the Atmospheric Impact Research Program, North Carolina State University, Raleigh, North Carolina.